DYNAMICS OF THE TOURISM INDUSTRY

Post-Pandemic and Post-Disaster Perspectives and Strategies

Advances in Hospitality and Tourism Series

DYNAMICS OF THE TOURISM INDUSTRY

Post-Pandemic and Post-Disaster Perspectives and Strategies

Edited by

Gül Erkol Bayram, PhD

Anukrati Sharma, PhD

First edition published 2024

Apple Academic Press Inc.
1265 Goldenrod Circle, NE,
Palm Bay, FL 32905 USA

760 Laurentian Drive, Unit 19,
Burlington, ON L7N 0A4, CANADA

CRC Press
2385 NW Executive Center Drive,
Suite 320, Boca Raton FL 33431

4 Park Square, Milton Park,
Abingdon, Oxon, OX14 4RN UK

Library and Archives Canada Cataloguing in Publication

Title: Dynamics of the tourism industry : post-pandemic and post-disaster perspectives and strategies / edited by Gül Erkol Bayram, PhD, Anukrati Sharma, PhD.
Names: Erkol Bayram, Gül, 1987- editor. | Sharma, Anukrati, 1981- editor.
Description: First edition. | Series statement: Advances in hospitality and tourism book series | Includes bibliographical references and index.
Identifiers: Canadiana (print) 20230554946 | Canadiana (ebook) 20230554970 | ISBN 9781774914489 (hardcover) | ISBN 9781774914496 (softcover) | ISBN 9781003455431 (ebook)
Subjects: LCSH: Tourism. | LCSH: Pandemics. | LCSH: Natural disasters.
Classification: LCC G155.A1 D94 2024 | DDC 338.4/791—dc23

Library of Congress Cataloging-in-Publication Data

CIP data on file with US Library of Congress

ISBN: 978-1-77491-448-9 (hbk)
ISBN: 978-1-77491-449-6 (pbk)
ISBN: 978-1-00345-543-1 (ebk)

About the Advances in Hospitality and Tourism Book Series

Editor-in-Chief:
Mahmood A. Khan, PhD
Professor, Department of Hospitality and Tourism Management,
Pamplin College of Business,
Virginia Polytechnic Institute and State University,
Falls Church, Virginia, USA
Email: mahmood@vt.edu

This series reports on research developments and advances in the rapidly growing area of hospitality and tourism. Each volume in this series presents state-of-the-art information on a specialized topic of current interest. These one-of-a-kind publications are valuable resources for academia as well as for professionals in the industrial sector.

BOOKS IN THE SERIES:

Food Safety: Researching the Hazard in Hazardous Foods
Editors: Barbara Almanza, PhD, RD, and Richard Ghiselli, PhD

Strategic Winery Tourism and Management: Building Competitive Winery Tourism and Winery Management Strategy
Editor: Kyuho Lee, PhD

Sustainability, Social Responsibility and Innovations in the Hospitality Industry
Editor: H. G. Parsa, PhD
Consulting Editor: Vivaja "Vi" Naraparedddy, PhD
Associate Editors: SooCheong (Shawn) Jang, PhD,
Marival Segarra-Oña, PhD, and Rachel J. C. Chen, PhD, CHE

Managing Sustainability in the Hospitality and Tourism Industry: Paradigms and Directions for the Future
Editor: Vinnie Jauhari, PhD

Management Science in Hospitality and Tourism: Theory, Practice, and Applications
Editors: Muzaffer Uysal, PhD, Zvi Schwartz, PhD, and
Ercan Sirakaya-Turk, PhD

Tourism in Central Asia: Issues and Challenges
Editors: Kemal Kantarci, PhD, Muzaffer Uysal, PhD, and Vincent Magnini, PhD

Poverty Alleviation through Tourism Development: A Comprehensive and Integrated Approach
Robertico Croes, PhD, and Manuel Rivera, PhD

Chinese Outbound Tourism 2.0
Editor: Xiang (Robert) Li, PhD

Hospitality Marketing and Consumer Behavior: Creating Memorable Experiences
Editor: Vinnie Jauhari, PhD

Women and Travel: Historical and Contemporary Perspectives
Editors: Catheryn Khoo-Lattimore, PhD, and Erica Wilson, PhD

Wilderness of Wildlife Tourism
Editor: Johra Kayeser Fatima, PhD

Medical Tourism and Wellness: Hospitality Bridging Healthcare (H_2H)©
Editor: Frederick J. DeMicco, PhD, RD

Sustainable Viticulture: The Vines and Wines of Burgundy
Claude Chapuis

The Indian Hospitality Industry: Dynamics and Future Trends
Editors: Sandeep Munjal and Sudhanshu Bhushan

Tourism Development and Destination Branding through Content Marketing Strategies and Social Media
Editor: Anukrati Sharma, PhD

Evolving Paradigms in Tourism and Hospitality in Developing Countries: A Case Study of India
Editors: Bindi Varghese, PhD

The Hospitality and Tourism Industry in China: New Growth, Trends, and Developments
Editors: Jinlin Zhao, PhD

Labor in Tourism and Hospitality Industry: Skills, Ethics, Issues, and Rights
Abdallah M. Elshaer, PhD, and Asmaa M. Marzouk, PhD

Sustainable Tourism Development: Futuristic Approaches
Editor: Anukrati Sharma, PhD

Tourism in Turkey: A Comprehensive Overview and Analysis for Sustainable Alternative Tourism
Editor: Ahmet Salih İkiz

Capacity Building Through Heritage Tourism: An International Perspective
Editor: Surabhi Srivastava, PhD

Medical Travel Brand Management: Success Strategies for Hospitality Bridging Healthcare (H2H)
Editors: Frederick J. DeMicco, PhD, RD, and Ali A. Poorani, PhD

Strategic Management for the Hospitality and Tourism Industry: Developing a Competitive Advantage
Editor: Vincent Sabourin, PhD

Tourist Behavior: Past, Present, and Future
Editors: Narendra Kumar, PhD, Bruno Barbosa Sousa, PhD, and Swati Sharma, PhD

The Hospitality and Tourism Industry in ASEAN and East Asian Destinations: New Growth, Trends, and Developments
Editors: Jinlin Zhao, PhD, Lianping Ren, D.HTM, and Xiangping Li, PhD

Brand Co-Creation Tourism Research: Contemporary Issues and Challenges
Editor: Raouf Ahmad Rather, PhD

Small Island and Small Destination Tourism: Overcoming the Smallness Barrier for Economic Growth and Tourism Competitiveness
Robertico Croes, PhD

Event Tourism in Asian Countries: Challenges and Prospects
Editor: Shruti Arora, PhD

Post-COVID Tourism and Hospitality Dynamics: Recovery, Revival, and Re-start
Editors: Umendra Narayan Shukla, PhD, and Sharad Kumar Kulshreshtha, PhD

Dynamics of the Tourism Industry: Post-Pandemic and Post-Disaster Perspectives and Strategies
Editors: Gül Erkol Bayram, PhD, and Anukrati Sharma, PhD

Medical, Dental, and Wellness Tourism: A Post-Pandemic Perspective
Mary Schreiber Swenson, PhD, and Amit Bansal

Dynamics of Heritage Management and Tourism Development: Trends, Practices, and Approaches
Editors: Surabhi Srivastava, PhD, and Ruchika Kulshrestha, PhD

ABOUT THE SERIES EDITOR

Mahmood A. Khan, PhD
Professor, Department of Hospitality and Tourism Management,
Pamplin College of Business, Virginia Tech's National Capital Region campus

Mahmood A. Khan, PhD, is a Professor in the Department of Hospitality and Tourism Management, Pamplin College of Business at Virginia Tech's National Capital Region campus. He has served in teaching, research, and administrative positions for the past 35 years, working at major U.S. universities. Dr. Khan is the author of several books and has traveled extensively for teaching and consulting on management issues and franchising. He has been invited by national and international corporations to serve as a speaker, keynote speaker, and seminar presenter on different topics related to franchising and services management. Dr. Khan has received the Steven Fletcher Award for his outstanding contribution to hospitality education and research. He is also a recipient of the John Wiley & Sons Award for lifetime contribution to outstanding research and scholarship; the Donald K. Tressler Award for scholarship; and the Cesar Ritz Award for scholarly contribution. He also received the Outstanding Doctoral Faculty Award from Pamplin College of Business. He has served on the Board of Governors of the Educational Foundation of the International Franchise Association, on the Board of Directors of the Virginia Hospitality and Tourism Association, as a trustee of the International College of Hospitality Management, and as a trustee on the Foundation of the Hospitality Sales and Marketing Association's International Association. He is also a member of several professional associations.

About the Editors

Gül Erkol Bayram, PhD
Associate Professor, School of Tourism and Hotel Management, Department of Tour Guiding, University of Sinop, Turkey

Gül Erkol Bayram, PhD, is currently working as an Associate Professor in the School of Tourism and Hotel Management, Department of Tour Guiding, University of Sinop, Turkey. Dr. Erkol Bayram has worked as an internal trainer in the tour guiding arena. Her doctorate is in Tourism Management from the Sakarya University, Turkey, and she completed her dissertation research on tour guiding in Turkey. Her core subjects are tourism, tour guiding, tourism policy and planning, women studies. She has edited a book on *Women and Tourism and Tour Guiding—Past, Present, Future* with Detay Publishing, Turkey. Currently, she is working on her book chapters under *Resilient and Sustainable Destinations after Disaster: Challenges and Strategies* under Emerald Publishing. She also has many book chapters under Cambridge, Emerald, CABI, IGI Global. She has been invited for many talks/lectures/panel discussions by different universities like University of Mumbai, India, International Multidisciplinary Conference, College for Women, Parade Ground, Jammu, India; International Faculty Development Programme, International Digital Conference 2021 on Curriculum 4.0., Jagdish Sheth School of Management and University of Management and Technology, Lahore, Pakistan; Alhamd Conference on Education, Islamic University, Islamabad, Pakistan; 11th Global Virtual Conference on Multidisciplinary Trends in Academic Research, Lead Philippines, Canada; The Mega Virtual International Conference 2021 on Multidisciplinary Hybrid Learning in Achieving Sustainable development Goals in Indonesia, etc. The Government of India and University of Kota invited her to give a lecture on tour guiding and making a career in tourism. She was invited as special guest, guest of honor, etc. to many conferences and organizations. She has recieved awards from several organizations. Dr. Bayram has also worked as a professional tour guide in Turkey.

Anukrati Sharma, PhD
Head and Associate Professor, Department of Commerce and Management, University of Kota, Rajasthan, India

Anukrati Sharma, PhD, is the Head and Associate Professor of the Department of Commerce and Management at the University of Kota (a State Govt. University) in Kota, Rajasthan, India. She is the Director of the Skill Development Centre of the same university. She is also Dean (honorary) of two faculties, Tourism and Hospitality and Aviation and Aerospace at Rajasthan Skill University (a Govt. State University) in Jaipur. In 2015, she received a research award from the University Grants Commission (UGC), New Delhi, for her project, Analysis of the Status of Tourism in Hadoti and Shekhawati Region/Circuit (Rajasthan): Opportunities, Challenges, and Future Prospects. Her doctorate from the University of Rajasthan is in Tourism Marketing, and she completed her dissertation research on Tourism in Rajasthan—Progress & Prospects. She has two postgraduate degree with specialties—in International Business (Master of International Business) and in Business Administration (Master of Commerce). Her interest areas are tourism, tourism marketing, strategic management, and international business management. Dr. Sharma is a featured author of Routledge. She is the Book Series Editor of *Building the Future of Tourism* under Emerald Publishing, UK. She has edited many books on tourism among which are Opportunities and Ventures, on Maximizing Business Performance and Efficiency Through Intelligent Systems, under IGI Global (Scopus indexed); Sustainable Tourism Development: Futuristic Approaches; Event Tourism in Asian Countries: Challenges and Prospects; and Event Tourism and Sustainable Community Development: Advances, Effects, and Implications, published Apple Academic Press (CRC Press, Taylor and Francis Group), USA, within the series Advances in Hospitality and Tourism; Tourism Events in Asia: Marketing and Development, under the book series Routledge Advances in Events Research; and Sustainable Destination Branding and Marketing: Strategies for Tourism Development, under CABI, UK. Others are Future of Tourism: An Asian Perspective, under Springer, Singapore; Overtourism as Destination Risk: Impacts and Solutions, under Emerald Publishing, UK, within the Tourism Security-Safety and Post Conflict Destinations book series; Overtourism, Technology Solutions, and Decimated Destinations, under Springer, Singapore; and The Emerald Handbook of ICT in Tourism and Hospitality, under Emerald Publishing, UK. She has also authored a book titled Event Management and Marketing Theory: Practical Approaches and Planning. Another book she wrote is titled International Best Practice in

Event Management, published by United Kingdom Event Industry Academy Ltd. and Prasetiya Mulya Publishing, Indonesia. Her current projects include editing books titled COVID-19 and Tourism Sustainability: Ethics, Responsibilities, Challenges, and New Directions for Routledge, USA; Festivals and Event Tourism: Building Resilience and Promoting Sustainability for CABI, UK; The Emerald Handbook of Destination Recovery in Tourism and Hospitality, Emerald Publishing, UK; and COVID-19 and the Tourism Industry: Sustainability, Resilience, and New Directions, Routledge, UK. She is also working on a major research project under the Mahatma Gandhi National Council of Rural Education, Ministry of Human Resource Development, Government of India.

A member of 10 professional bodies, Dr. Sharma has attended a number of national and international conferences and presented 45 papers. She has been invited as keynote speaker and panel member by different countries such as Sri Lanka, Uzbekistan, Nepal, and Turkey. She has also been invited as a visiting professor to Kazakhstan and Uzbekistan.

Contents

Contributors

Sumedha Agarwal
School of Business Studies, Sharda University, Noida, India

Alhamzah Alnoor
School of Management, MBA, Organizations Studies, Universiti Sains Malaysia, Malaysia

Manpreet Arora
School of Commerce and Management Studies, Central University of Himachal Pradesh, Dharamshala, India

Andi Asrifan
English Language Department, Universitas Muhammadiyah Sidenreng Rappang, Indonesia

Fatmanur Kubra Aylan
Recreation Management Department, Tourism Faculty, Selçuk University, Konya, Turkey

Ahu Yazici Ayyildiz
Faculty of Tourism, Aydin Adnan Menderes University, Kusadasi, Aydin, Turkey

Gokcenur Avci
Department of Tour Guiding, School of Tourism and Hospitality Management, Sinop University, Sinop, Turkey

Gül Erkol Bayram
Department of Guiding, School of Tourism and Hospitality,, University of Sinop, Sinop, Turkey

Ali Turan Bayram
Department of Tour Guiding,, School of Tourism and Hospitality Management, Sinop University, Sinop, Turkey

Rohan Bhalla
Department of Tourism and Hospitality Management, Jamia Millia Islamia, New Delhi, India

Halide Bilgin
Institution of Social Sciences, Anadolu University, Eskişehir, Turkey

Maria Boaler
School of Management Studies, Reva University, Bangalore, India

Yavuz Cetın
Institution of Social Sciences, Tourism Management, Aydin Adnan Menderes University, Aydin, Turkey

Abdullah Eravcı
Boyabat Vocational Higher School, Member of Department Wholesale and Retail, Sinop University, Sinop, Turkey

Aslı Sultan Eren
Intuition of Tourism Research, Tour Guiding, Nevşehir Hacı Bektaş Veli Universty, Turkey

Seda Erkekli
Faculty of Tourism, Recreation Management, Ankara Hacı Bayram Veli Universty, Ankara, Turkey

Hatice Sari Gok
Department of Travel, Tourism and Entertainment Services, Yalvaç Vocational School, Isparta University of Applied Science, Isparta, Turkey

Kubra Gundogdu
Instution of Social Sciences, Tourism Management, Selçuk Universty, Konya, Turkey

Hasan Tahsin Kavlak
Faculty of Tourism, Recreation Management, Ankara Hacı Bayram Veli Universty, Ankara, Turkey

Maximiliano E. Korstanje
Departments of Economic, University of Palermo, Mario Bravo, Buenos Aires, Argentina

Miray Kose
School of Tourism and Hospitality Management, Department of Tour Guiding, Sinop University, Sinop, Turkey

Havvanur Kum
Institution of Social Sciences, Anadolu University, Eskişehir, Turkey

Jeetesh Kumar
School of Hospitality, Tourism and Events; Centre for Research and Innovation in Tourism (CRiT); Sustainable Tourism Impact Lab, Taylor's University, Malaysia

Enkeleda Lulaj
Faculty of Management in Tourism Hospitality and Environment, Haxhi Zeka University, Peja, Kosovo

Muthmainnah
Indonesian Language Department, Universitas Al Asyariah Mandar, Teacher Training and Education Faculty, Indonesia

Farhad Nazir
University of Coimbra, CEGOT, Faculty of Arts & Humanities, Portugal University of Swat, Pakistan

Nazik Çelikkanat Pasli
Faculty of Tourism, Department of Recreational Management, University of Giresun, Giresun, Turkey

Neethiahnanthan Ari Ragavan
Faculty of Social Sciences & Leisure Management, Taylor's University, Malaysia

Melike Sak
Instution of Social Sciences, Travel Management and Tourism Guiding, Selcuk University, Konya, Turkey

Sofia Devi Shamurailatpam
Department of Banking and Insurance, The Maharaja Sayajirao University of Baroda, Vadodara, Gujarat, India

Anukrati Sharma
Department of Commerce and Management, University of Kota, Kota, Rajasthan, India

Roshan Lal Sharma
Central University of Himachal Pradesh, Dharamshala, Himachal Pradesh, India

Parag Sunil Shukla
Department of Commerce and Business Management, The Maharaja Sayajirao University of Baroda, Vadodara, Gujarat, India

Priya Singh
Department of Tourism and Hospitality Management, Jamia Millia Islamia, New Delhi, India

Georgia Tacu
Gh. Zane' Institute for Economic and Social Research Iași, Romanian Academy, Romania

Abdul Wafi
Institut Agama Islam Negeri (IAIN) Madura, Tarbiyah Faculty, Indonesia

Author Profiles

Nazik Çelikkanat Pasli is currently teaching at the Department of Recreational Management, Faculty of Tourism at Giresun University. She has completed her doctorate studies in the field of recreational management at Ankara Haci Bayram Veli University. She has conducted various studies in the fields of tourism and recreation.

Melike Sak completed her undergraduate education at Sinop University, School of Tourism and Hotel Management, Tourist Guidance; the author is doing her master's degree in Selcuk University, Institute of Social Sciences, Travel Management and Tourism Guidance Department. The author has presented national and international papers in various tourism fields. She has an international book chapter and national and international papers and articles. The author has various studies in the field of sustainable tourism, health tourism, cultural heritage tourism, ecotourism, event tourism, sustainable tourism, event tourism, and tourism guidance, and she has participated in national and international congress meetings and seminars.

Muthmainnah is an Assistant Professor at the Universitas Al Asyariah Mandar in West Sulawesi, Indonesia. She is a lecturer, a global speaker, a researcher, and an international leader. Some of her achievements at her university are as follows: she was the Chairman of the Indonesian Language Department at the Teacher Training and Education Faculty, she was in public relations department in her university, she was the Director of Women's Centre Studies, and also served as Deputy Director of the Quality Assurance Unit at her university. She is now the Deputy Director of the Language and Character Development Institute at her university. Currently, she is active as an international member, international conference coordinator, and international trainer. She is also on the International Board for Professors and Experts of the Scientific Innovation Research Group, Egypt (www.sirg.club), and the International Board of TEFL Kuwait (Member at Large) representing Asia. She is also a country head, an ambassador, an international reviewer, an editor board, and an international advisory board of international journals, and she is also a country director for many international organizations representing Indonesia. She has excellent experience as a visiting professor lecturing on ICT, technology in education, research methods, education management, empowerment of women, entrepreneurship, teaching, research and development, and many other topics relevant to ICT research and teacher development, SDGS for quality education and quality enhancement in higher education at many universities.

She is the winner of more than 60 awards including Outstanding Professor, Outstanding Leadership, Best International Influencer, Best Emerging Professor of the Year, International Award for Working Women, Best Speaker, SDGS Warrior, Best Global Outstanding Educational Innovation, etc. Her international achievements are as Doctor Honouris Causa from Layahe University and Candidate Doctor HC from the Philippines. She is also a member of the United Nations Volunteer Roster. She is on the Advisory Board of Indonesian Education Share to Care Volunteers. Her interests and her doctorate areas are developing instructional materials, ASSURE, TEFL, ICT in education and cybergogy. She is ready for international collaboration and teamwork.

Andi Asrifan, SPd, MPd is a lecturer in the Faculty of Teachers Training and Education, Universitas Muhammadiyah Sidenreng Rappang (UMS Rappang) in South Sulawesi, Indonesia. He is also the Head of Students and Alumni Affairs of Universitas Muhammadiyah Sidenreng Rappang (UMS Rappang). He received his PhD (Language Education, Concentration in English Language) and MPd (Master Pendidikan/Master of Education) from the State University of Makassar, South Sulawesi, Indonesia in 2018 and 2011, respectively. He received his Bachelor's degree (SPd) in Faculty of Teachers Training and Education from Universitas Muhammadiyah Parepare, South Sulawesi, Indonesia, in 2007. He got a sandwich-model scholarship at Northern Illinois University, Dekalb, IL, USA, in 2015. He has published some research articles in leading journals and conference proceedings, including Scopus and Thomson Reuters. His research interest is in English Language Teaching (ELT), ESP, linguistics, and technology in education. He has full membership of Himpunan Penerjemah Indonesia/Association of Indonesian Translator (HPI-01-19-3105). He is a Member of Laureate—Cambridge Online Language Learning Research Network (OLLReN). In 2012–2016, he was the Head of the Educational Technology Department of STKIP Muhammadiyah Rappang. He was a Director of the Center for Language and Indonesian Culture of STKIP Muhammadiyah Rappang in 2012–2015. He was the Head of the International Office of STKIP Muhammadiyah Rappang in 2015–2016. In 2016–2019, he was elected to be the vice-chancellor in finance and human resources in STKIP

Muhammadiyah Rappang. He is an editorial board member and reviewer in several international and national journals.

Abdul Wafi is an Assistant Professor at IAIN Madura. He is a researcher, educationist, and international speaker. He earned his undergraduate program in English Letters at University of Pesantren Darul Ulum (UNIPDU) Jombang in 2006 and pursued his postgraduate program in English Language Teaching (ELT) at State University of Malang in 2011. He gained experienced as an English Trainer at Mahesa Institute Pare—Kediri, known as Kampung Inggris, from 2002 to 2009. He is the author and master trainer of Terapi Ngomong Inggris Versi MUI (English Speaking Therapy, MUI Version); the CEO of SMiLE—The Cram English course and Training Pamekasan – Madura. He is also as one of editors in re-JIEM *Journal MPI IAIN Madura* and *JETLEE (Journal of English Teaching, Linguistics, and Literature)* IAIN Lhokseumawe. His interests are ELT, character education, english literature, linguistics, Islamic teachings, leadership, Islamic educational management, and education for early children. He wrote many articles related to his expertise and interest—Creative, Attractive, Imaginative, and Joyful English Teaching (CRAIJET). During the COVID-19 pandemic, he was actively involved in various virtual events such as conference, webinar, annual summit as speaker and or as organizer. He organized several virtual events such as lecture talk, APDEIT International Conference, Virtual Students Conference (VISCON), Virtual International Conference (VIC). He is a member of ELITE—Association of English Lecturer under PTKIN Indonesia with Reg. Number: 000701. He is a Dept. Protocol and Scientific Events of Indonesian Education Share-to-Care Volunteer (IES2C); as state coordinator of Charles Walter's Society for Innovation & Research (CWSIR); International Ambassador of Startup Community—Bangladesh; a Global Ambassador of Dynamic Peace Rescue Mission International—Nigeria; as Icon of World Peace for Brazilian Association Human Rights; a Honorary High Degree for Association of Charity Permata Ikhlas Sabah—Malaysia; a Rising Star of the Year 2021 for KGN Humanity Social Services Foundation; an International Ambassador of Peace for World Literary Forum for Peace and Human Rights (WLFPH). He holds a life time membership of Cape Forum Youth Trust—India; lifetime membership of Institute Global Professional (IGP) Bangladesh with ID: IGP01CvCRvBD115529; a professional membership of EDUKOS Unite of Scholars; a membership of Impact Youth Sustainability; a membership of National Institute for Technical Training & Skill Development with member ID: NITTSD/PROFESSOR/00551. He was awarded a Doctorate of Philosophy in global development by Theophany International University and Leaders Autonomy International.

Hatice Sari Gök is currently an Associate Professor in Yalvaç Vocational School, Department of Travel, Tourism, and Entertainment Services at the Isparta University of Applied Sciences, Isparta, Turkey. She completed her masters and undergraduate studies at Gazi University. She received her language education in travel management and tourist guidance teaching program. She has also worked in the reservation department of the hotel business in the tourism sector, in the front office department. Sari Gök has many academic publications (articles, conference proceedings, book chapters) related to tourism. Her research areas include community-based tourism, destination management, tourism marketing, tourism types, and gastronomy. She also serves on the referee board of several national and international journals.

Fatmanur Kübra Aylan is an Associate Professor in the Department of Recreation Management in Faculty of Tourism at the Selçuk University, Konya, Turkey. She received her PhD and MS in tourism management at Selçuk University and Bachelor's degree in hospitality management teaching at Nevşehir Haci Bektaş Veli University. Dr. Aylan also worked as receptionist in tourism sector before working at the university. Dr. Aylan has many academic publications (articles, conference papers, book chapters) related to tourism. Her research areas include responsible tourism, recreation management, and destination marketing. She also serves on the board of referees for several national and international journals.

Kübra Gündoğdu completed her undergraduate education in Sinop University, School of Tourism and Hotel Management, Accommodation Management and is doing her master's degree in Selçuk University, Institute of Social Science, Tourism Management Department. The author has international papers and articles on sustainability and voluntary tourism. She made a presentation at the International Congress. Her areas of interest are sustainable tourism and tourism types that develop depending on the understanding of sustainability.

Anukrati Sharma, PhD, is the Head and Associate Professor of the Department of Commerce and Management at the University of Kota (a State Govt. University) in Kota, Rajasthan, India. She is the Director of the Skill Development Centre of the same university. She is also Dean (honorary) of two faculties, Tourism and Hospitality and Aviation and Aerospace at

Rajasthan Skill University (a Govt. State University) in Jaipur. In 2015, she received a research award from the University Grants Commission (UGC), New Delhi, for her project, Analysis of the Status of Tourism in Hadoti and Shekhawati Region/Circuit (Rajasthan): Opportunities, Challenges, and Future Prospects. Her doctorate from the University of Rajasthan is in Tourism Marketing, and she completed her dissertation research on Tourism in Rajasthan—Progress & Prospects. She has two postgraduate degree with specialties— in International Business (Master of International Business) and in Business Administration (Master of Commerce). Her interest areas are tourism, tourism marketing, strategic management, and international business management. Dr. Sharma is a featured author of Routledge. She is the Book Series Editor of *Building the Future of Tourism* under Emerald Publishing, UK. She has edited many books on tourism among which are—Opportunities and Ventures, on Maximizing Business Performance and Efficiency Through Intelligent Systems, under IGI Global (Scopus Indexed); Tourism Sustainable Tourism Development Futuristic Approaches under Apple Academic Press (CRC Press, Taylor and Francis Group), USA, within the series Advances in Hospitality and Tourism; and a book titled *Tourism Events in Asia Marketing and Development Routledge*, USA, under Advances in Events Research Series; Sustainable Destination Branding, and Marketing: Strategies for Tourism Development under CABI, UK. Others are Future of Tourism: An Asian Perspective, under Springer, Singapore; Over-tourism as Destination Risk: Impacts and Solutions under Emerald Publishing, UK, within the Tourism Security—Safety series; Over-Tourism, Technology Solutions, and Decimated Destinations, under Springer, Singapore; Event Tourism in Asian Countries: Challenges and Prospects under Apple Academic Press (Taylor and Francis Group), USA; and The Emerald Handbook of ICT in Tourism and Hospitality, under Emerald Publishing, UK. She has also authored a book titled *Event Management and Marketing Theory, Practical Approaches and Planning.* Another book she wrote is titled *International Best Practice in Event Management* and it was published by United Kingdom Event Industry Academy Ltd. and Prasetiya Mulya Publishing, Indonesia. Her current projects include editing a book on *COVID-19 and Tourism Sustainability: Ethics, Responsibilities, Challenges, and New Directions* for Routledge, USA; *Festivals and Event Tourism: Building Resilience and Promoting Sustainability* for CABI, UK; *The Emerald Handbook of Destination Recovery in Tourism and Hospitality*, Emerald Publishing, UK, COVID-19 and the *Tourism Industry: Sustainability, Resilience, and New Directions*, Routledge, UK. She is also working on a major research project under Mahatma Gandhi National Council of Rural Education, Ministry of Human Resource Development, Government of India. A member of 10 professional bodies, she has attended a number of national and international conferences and presented 45 papers. She has been invited as keynote speaker and panel member by different countries such as Sri Lanka, Uzbekistan, Nepal, and Turkey. She has also been invited as a visiting professor to Kazakhstan and Uzbekistan.

Gül Erkol Bayram is currently an Associate Professor in the School of Tourism and Hotel Management, Department of Tour Guiding, University of Sinop, Sinop, Turkey. Her doctorate is in tourism management from the Sakarya University, Turkey, and she completed her dissertation research on tour guiding in Turkey. Her core subjects are tourism, tour guiding, tourism policy and planning, and women studies. Dr. Bayram has also worked as a professional tour guide in tourism sector. She has many book chapters and articles in national and international arena. She has been invited for many talks/lectures/panel discussions by different universities.

Asli Sultan Eren completed her undergraduate education at Sinop University, School of Tourism Management and Hotel Management, Department of Tourist Guidance and her master's degree at Nevşehir Hacı Bektaş Veli University, Institute of Tourism Research, Department of Tourism Guidance. As a PhD Candidate, she is still continuing her education at Nevşehir Hacı Bektaş Veli University, Department of Tourism Guidance. She has presented national and international papers in various tourism fields. The author has an international book chapter, national and international papers and articles. Her research interests include tour guiding, tourist behaviour, tourism ethics, gastronomy tourism, cultural heritage tourism and event tourism.

Alhamzah Alnoor is a professional administrator with 10 years of experience in organizational studies, social commerce, internship programs, multi-criteria decision analysis, leadership and innovation, strategic planning, and technology acceptance models. He has successfully achieved several projects during his career with impactful business values. Creative, flexible, motivated with active optimism and belief in diversity and inclusion. He is a reviewer for many journals. He published many papers in different and high-impact journals. He is a senior lecturer at the Southern Technical University, Management Technical College. He received his MBA from the University of Basrah, Iraq. Now, he is a PhD scholar at the School of Management, Universiti Sains Malaysia, Malaysia.

Farhad Nazir is a Lecturer at Institute of Cultural Heritage, Tourism, and Hospitality Management, University of Swat, Pakistan. Before joining University of Swat, he has served as a coordinator at Marriott International Hotels Islamabad, Pakistan. He has an MPhil degree in Development Studies from Pakistan Institute of Development Economics, Islamabad. His master's degree is in Tourism and Hospitality from Hazara University, Pakistan, with distinction. His research lines address cultural tourism and cultural identity, Islamic legislation and tourism laws, and hospitality laws. He has authored several publications in recognized journals of Pakistan. He has supervised 20 graduate students. Also, he has participated in a number of national conferences. During his career at the Marriott Hotel Islamabad, he executed various training sessions for employees as master trainer.

Jeetesh Kumar is a Senior Lecturer at School of Hospitality, Tourism, and Events; Associate Director for Information Management & Documentation at Centre for Research, and Innovation in Tourism (CRiT), Taylor's University, Malaysia. His doctorate is from the Taylor's University in hospitality and tourism, with research on economic impacts of business events in Malaysia. He has two postgraduate degrees specializing in hospitality management and international tourism from University of Toulouse, France and the other in business administration (MBA—Marketing) from Hamdard University, Pakistan. His research areas include economic impacts, economic modeling, MICE, medical tourism, and behavioral studies. He has worked on consultancy and research projects at the national/international level and authored 45 publications including research articles and book chapters. A total of 15 postgraduate students have graduated under his supervision, and currently he is supervising seven PhD scholars. Along with being an active member of several national and international associations, conferences, and journals, he is also serving as an Associate Editor for *Asia-Pacific Journal Innovation in Hospitality and Tourism (APJIHT).*

Neethiahnanthan Ari Ragavan is Professor and the Executive Dean, Faculty of Social Sciences and Leisure Management, Taylor's University, Malaysia. He oversees five schools and one institute, among which one of the schools is currently ranked amongst top 20 in the world by QS World University Rankings by Subject which is the School of Hospitality, Tourism, and Events and it is internationally benchmarked by the United Nations World Tourism Organization (UNWTO) Ted Qual certification. He has over 27 years of academic experience and holds a doctorate degree from France. He is majorly involved in hospitality and tourism education in Malaysia and the region. In November 2013, he initiated the setting-up of the ASEAN Tourism Research Network or now aptly known as ASEAN Tourism Research Association (ATRA), a regional cooperation among universities in ASEAN countries for greater research and education collaboration on issues related to hospitality and tourism. He is presently the President of ASEAN Tourism Research Association (ATRA) and has been active in various international boards such as Pacific Asia Travel Association (PATA). His areas of research include on higher education models and system, hotel and tourism development and innovation; and, food and culture.

Hasan Tahsin Kavlak is currently a Research Assistant in Faculty of Tourism, Department of Recreation Management, University of Ankara Haci Bayram Veli University, Ankara, Turkey. His master's degree is in recreation management from the Gazi University, Turkey, and he completed his dissertation research on multiple intelligence and recreation in Turkey. His doctorate dissertation research continues in the field of destination management. His core subjects are recreation, recreation psychology, and tourism marketing. He has many book chapters and articles in national and international arena.

Seda Erkekli completed her undergraduate education in Gazi University Faculty of Tourism, Department of Recreation Management, works at Ankara Haci Bayram Veli University. Currently, she is continuing her doctorate in Ankara Haci Bayram Veli University Graduate Education Institute, Department of Recreation Management. The author has national and international papers and articles. She has studies in fields such as management, recreation, and tourism.

Priya Singh is presently working as an Assistant Professor, hotel management in Jamia Millia Islamia, New Delhi. She is a doctorate in hotel management and possesses more than 10 years of academic teaching and industry experience at various reputed organizations like IHM Kurukshetra, IHM Dehradun, and The Oberoi Amarvilas, Agra. She has also qualified UGC-NET/JRF in tourism management. She is a budding and passionate researcher, and her core areas of research are hospitality education, culinary tourism, and social media food trends. She has presented research papers at various national and international conferences. She has numerous publications in reputed journals and books. She is also a Certified Hospitality Trainer (CHT), approved by the ministry of tourism. She has expertise in curriculum development and has also designed various skill development programs for different sectors.

Sumedha Agarwal is a Junior Research Fellow pursuing her PhD from the School of Business Studies, Sharda University, Noida, India. She has done masters in tourism management from IMS Ghaziabad and holds a master degree in conservation, preservation, and heritage management from Indraprastha University, Delhi. She has teaching experience of five years. She has worked with prestigious institutions like the Department of Tourism and Hospitality Management, Jamia Millia Islamia, New Delhi and Amity University Noida, India. She has one edited book to her credit and 12 publications in national and international journals, books and edited books. She has presented her research works at several international conferences, including one attended at the University of Kragujevac, Serbia. Presently, she is working in the area of sustainable tourism, women entrepreneurship, community-based tourism, and homestays, mainly in the states of Uttarakhand and Himachal Pradesh

Rohan Bhalla is an SRF pursuing his PhD from the Department of Tourism and Hospitality Management, Faculty of Humanities and Languages, Jamia Millia Islamia (a Central University). He is a gold medalist in PGDM (International Business) from the Indian Institute of Tourism and Travel Management, an organization of Ministry of Tourism, Government of India and has completed BBA (Finance) from Jiwaji University Gwalior, India. He also completed an advanced research course in research from Mid-Sweden University, Sweden. He has academic experience of 3 years at all levels of imparting education, including his association with the Indian Institute of Management, Indore. He has nine publications to his credit in international and national journals, books and edited books. His area of expertise is qualitative research, and he works in a multi-disciplinary approach, particularly in real-time social sciences projects. His interest lies in philosophy of science, spirituality, transformational tourism, regenerative tourism, rural tourism, and gender studies. He also takes an interest in behavioral sciences, psychology, communication, and personality development programs.

Yazici Ayyildiz is currently an Associate Professor in Faculty of Tourism, University of Aydin Adnan Menderes, Aydin, Turkey. Her doctorate is in tourism management from the Aydin Adnan Menderes University, Turkey and she completed her dissertation research on private label usage and applications in retailing. Her core subjects are accommodation businesses and tourism marketing. She has conducted research on marketing, service marketing, and customer relationship management within the basic area of social, humanities, and administrative sciences. Dr. Ayyildiz has many book chapters and articles in national and international arena. Dr Yazici Ayyildiz has been invited by many universities and tourism businesses for many speeches/conferences/panel discussions and has also advised many graduate and doctoral students and continues her studies meticulously.

Yavuz Çetin, who completed his undergraduate education at Sinop University School of Tourism and Hotel Management with a certificate of honor as the top student in the school, continues his graduate education at Aydin Adnan Menderes University, Social Sciences Institute, Department of Tourism Management. The author has papers and articles on national and international platforms. He has studies on various subjects such as responsible tourism, gastronomy tourism, ecotourism, tourism promotion, tourism and destination marketing. The author has participated and presented papers in various fields of tourism at national and international congress meetings and seminars.

Parag S. Shukla holds a PhD degree in commerce and business management with focus on strategic marketing in the area of "retailing." He has been working as an Assistant Professor in The Maharaja Sayajirao University of Baroda since 2009. He has presented and published many research papers in contemporary areas of marketing in national and international journals. He is also an author in a book titled Retail Shoppers' Behaviour in Brick and Mortar Stores—A Strategic Marketing Approach which is published by a reputed publisher. His major research areas of interest includes retailing, services marketing, and consumer behavior to name a few.

Sofia Devi Shamurailatpam has a PhD in economics with specialization in the area of banking and financial economics. Currently, she is serving as an assistant professor in the Department of Banking and Insurance, Faculty of Commerce, The Maharaja Sayajirao University of Baroda. She has published several research papers in her credit and authored a book titled Banking Reforms in India: Consolidation, Restructuring and Performance published by Palgrave Macmillan, UK (2017). Her major research areas of interest includes economics of banking, financial economics, economics of gender, agricultural economics, and development economics particularly contemporary issues on sustainability.

Maximiliano E. Korstanje is Editor-in-Chief of *International Journal of Safety and Security in Tourism* (UP Argentina) and Editor-in-Chief *Emeritus of International Journal of Cyber Warfare and Terrorism* (IGI-Global US). In 2015, he was awarded as Visiting Research Fellow at School of Sociology and Social Policy, University of Leeds, UK and the University of La Habana Cuba. In 2017, he was elected as Foreign Faculty Member of AMIT, Mexican Academy in the study of Tourism, which is the most prominent institution dedicated to tourism research in Mexico. He was also selected to take part in the 2018 Albert Nelson Marquis Lifetime Achievement Award, a great distinction given by Marquis Who´s Who in the world.

Ali Turan Bayram is currently an associate professor in School of Tourism and Hotel Management, Department of Tour Guiding, University of Sinop, Sinop, Turkey. His doctorate is in Tourism Management from the Gazi University, Turkey and he completed his dissertation research on e-service quality. His core subjects are tourism, tour guiding, alternative tourism, e-service, and technological changes in tourism. He has many book chapters and articles in national and international arena.

Miray Köse who is final year student at Sinop University, Department of Tourism Guidance, worked as an apprentice at the Gri Tour Travel Agency and also worked as an agency officer and hotel guide at the Ets Tur. Her key study subjects are tourism, tour guiding, and faith tourism.

Gökçenur Avci, who is final year student at Sinop University, Department of Tourism Guidance, worked as agency officer and hotel guide at the Ets Tur. Her key study subjects are tourism, tour guiding, and health tourism.

Halide Bilgin completed her undergraduate education at Sinop University School of Tourism and Hotel Management. She continued her graduate education at Anadolu University, Social Sciences Institute, Department of Tourism Management. She also went to Poland for one semester with Erasmus Exchange program. The author has papers and articles on national and international platforms. She has studies on various subjects such as sustainable tourism, ecotourism, tourism and destination marketing. The author has participated and presented papers in various fields of tourism at national and international congress meetings and seminars.

Havvanur Kum completed her undergraduate education at Sinop University School of Tourism and Hotel Management. She continued her graduate education at Anadolu University, Social Sciences Institute, Department of Tourism Management. The author has papers and articles on national and international platforms. He has studies on various subjects such as sustainable women studies, organizational behavior, tourism and destination marketing. The author has participated and presented papers in various fields of tourism at national and international congress meetings and seminars.

Enkeleda Lulaj is a Doctor of Science in Finance and Accounting (Economy), works as an Assistant Professor at the Faculty of Management in Tourism, Hospitality, and Environment, University "HaxhiZeka" Peja-State of Kosovo. She is the cofounder of B.O.R.N., Ambassador for Financial Evaluation, Ambassador for Women-Tech, as well as the Ambassador for Research and Innovation. She is the Country Director of International Young Society and Country Director of World Voice International. She is a member of the editorial board of many magazines in different countries of the world such as USA, Belgium, Georgia, Germany, Denmark, Ukraine, India, Pakistan, Philippines, etc. She has great experience in international conferences as a supervisor, as a member of the scientific committee, as a member of the editorial board, as a session chair in universities, institutes, and various countries of the world as follows: Belgium, France, Switzerland, Austria, Turkey, the United Kingdom, South Korea, India, Italy, etc. She is a reviewer in prestigious journals on Scopus and Web of Science such as: the United Kingdom and Switzerland. She also has an excellent experience as a visiting professor to lecture on economics, finance, accounting, education, teaching, research, and many other topics relevant to economic development, financial stability, and quality enhancement in higher education at many universities, institutes, NGOs, symposiums, summits, workshops, webinars around the world such as: the United States, India, Indonesia, Egypt, Philippines, Belgium, Germany, Albania, Italy, Iran, Australia, Iraq, Egypt, Malaysia, Romania, Ireland, Cambodia, England, Germany, France, Canada, Edinburgh, Hong Kong, Poland, etc. She has completed many trainings and programs in the field of economics, finance, accounting, research, management, diversity, etc., in universities worldwide, and is certified from the World Bank and the International Monetary Fund through the Edx program. She is a member of Institutes of Economics, Education, and Research. She has experience in projects through Erasmus +, Euphoria, Tempus programs. She is the winner of many awards in areas such as: teaching, science, creative ideas, a successful woman, peace, etc. In Kosovo, she has an excellent experience in many functions within the University as a member of the university board, senator, coordinator for academic development and quality assurance, member of the commission for internationalization, and member of many other committees within the faculty and university. She has been a distinguished student in Kosovo and now has a number of acknowledgments from students as a distinguished professor.

Maria Boaler is a researcher in Faculty of Commerce and Management Studies Reva University, Bangalore. Her research interest is in tourism, management, and related areas. She has publications in national and international concept.

Roshan Lal Sharma is a Professor and Chair, Department of English and Dean, School of Languages in Central University of Himachal Pradesh at Dharamshala. A Senior Fulbright Fellow at UW-Madison (USA) and IRH (Institute for research in the Humanities) Honorary Fellow during 2007–08, Sharma has authored *Shorter Fiction of Raja Rao* (Rawat, 2009), and *Walt Whitman* (Vrinda, 2000); co-authored *Som P. Ranchan: Dialogue Epic in Indian English Poetry* (Pencraft, 2012); and co-edited *Gendered Spaces and Ruptured Identities: Representation of Women in African Literature* (AuthorsPress, 2022), *Mapping Diaspora Identities* (Anamika, 2017), *Communication in Contemporary Scenario: Its Multiple Dimensions* (Anamika, 2017), *Communication, Entrepreneurship and Finance: Renegotiating Diverse Perspectives* (Anamika, 2018), *Envisioning Effective Management Communication* (Anamika, 2019), and *Communication Perspectives in Modern Businesses* (Anamika, 2020). He has more than 80 papers and book chapters to his credit published by reputed publishers such as Emerald (*Foresight*), TU, Dublin (*IJRTP*), CRC Press, Taylor and Francis, Routledge, Sage, Apple Press, Palgrave MacMillan, IGI International, *SHSS*, *Dialog*, *Indraprastha*, *JOA*, etc.

Manpreet Arora, PhD, is Senior Assistant Professor of Management in the School of Commerce and Management Studies, Central University of Himachal Pradesh Dharamshala, India. With around 20 years of teaching experience, she has varied interest areas. A gold medalist at undergraduate and distinction holder at postgraduate level, she obtained her PhD in international trade from Himachal Pradesh University, Shimla, India. She received her MPhil also from the same university. She is a throughout merit holder with meritorious ranks in 10th and 12th also. Her areas of research interest include accounting and finance, strategic management, entrepreneurship, qualitative research, case study development, communication skills and microfinance. She has been guiding research at the doctoral level and has worked in the area of microfinance. As an active seminarian, she has attended more than 100 seminars/conferences, and has visited several universities and colleges to deliver invited talks on finance, strategic management, qualitative research, business communication, interpersonal skills, entrepreneurship and skill development. She is a motivational speaker and conducts workshops on communication and motivation. Having published more than 25 papers in various journals of national and international repute (including SCOPUS,

WOS, and Category Journals), she has also worked as content developer of MHRD e-PG Pathshala Project and OERs for IGNOU. She has written thirty book chapters in national as well as international books/handbooks/volumes published with Routledge, CABI, Apple Academic Press, IGI, Taylor and Francis and latest with Springer Nature etc. With four edited books to her credit, she is a persistent researcher in the field of management. She is an active social worker also and is working towards protecting the rights of women.

Georgia Tacu is Researcher at "Gh. Zane" Institute for Economic and Social Research, Iași Branch of Romanian Academy. Her primary research interests are in cultural tourism and rural tourism. She is also a co-organizer of the International Conference on Tourism and Rural Space in national and international context - TARS. She has many papers, book chapters, and articles in rural tourism.

Abdullah Eravcı is currently an Assistant professor in the University of Sinop, Sinop, Turkey. His doctorate is in management science from Istanbul Gelişim University. His core subjects are destination branding and marketing, consumer behavior. He has many book chapters and articles in the national and international arena.

Abbreviations

ADTOI	Domestic Tour Operators of India
AHLA	American Hotel and Lodging Association
ASI	Alpine Safety and Information Centre
COMCEC	Committee for Economic and Commercial Cooperation
COVID-19	coronavirus disease 2019
CRS	Congressional Research Service
DMO	destination marketing organization
ECDC	East Central Development Corporation
EMA	European Medicine Agency
FSSAI	Food Safety and Standards Authority of India
GDP	gross domestic product
HIV/AIDS	human immunodeficiency virus/acquired immune deficiency syndrome
ICMR	Indian Council of Medical Research
MAH	Malaysian Association of Hotels
MERS	Middle East respiratory syndrome
MICE	meetings, incentives, conventions, and exhibitions
NABARD	The National Bank for Agriculture and Rural Development
NCAER	National Council of Applied Economic Research
NGO	non-governmental organization
NRAI	National Restaurant Association of India
OECD	Organization for Economic Co-Operation and Development
OEF	Oxford Economic Forecasting
PCR	polymerase chain reaction
QCI	Quality Council of India
SAATHİ	System for Assessment, Awareness, and Training for Hospitality Industry
SARS	severe acute respiratory syndrome
SHG	self-help group
SIDBI	Small Industries Development Bank of India
SME	small medium enterprises
SOP	standard operating procedures

TCI	tourism climate index
TCM	theory context method
TÜRSAB	Türkiye Seyahat Acenteleri Birligi
TUYED	Association of Tourism Writers and Journalist
UNCTAD	United Nations Conference on Trade and Development
UNDP	United Nations Development Programme
UNFCCC	United Nations Framework Convention on Climate Change
UNWTO	United Nations World Tourism Organization
WTTC	World Travel and Tourism Council
WWOOF	Willing Workers on Organic Farms
H1N1	swine flu
H5N1	avian influenza

Preface

Since ancient times, people have felt the need to travel for different purposes. The rapid development of the tourism sector, which started with the industrial revolution, has raised different issues such as transportation, service, and infrastructure, and has increased the need for people to travel. Changes in technology, workforce, economy, and sociocultural structure have played major roles in the improvement of tourism. Such developments around the world have increased with technology. Thanks especially to the Internet, people have visited, seen, and experienced different destinations across the globe. Tourism has always been renewed and redesigned with technology. Tourism has gained a more solid, static, and powerful position in the new century, with the changes in the tourist profile, differentiating needs and demands.

The rapid development of tourism until 2020 started to decline with the emergence of COVID-19 epidemic. The world faced an unprecedented global health pandemic in 2020. The COVID-19 pandemic, which emerged in Wuhan, China, spread all over the world in a very short time; many people lost their lives, existing workplaces were closed, and national and international tourism movements came to a standstill. In this process, a long way has been covered and short–medium- and long-term plans have been made, measures have been taken, and incentive laws have been enacted, both in the public context and by international organizations. Although the pandemic is losing its effect, it is not possible to say that such an epidemic will not drag us into such a big crisis in the future.

Another crisis faced by the tourism sector is disasters. Destructions or natural disasters that come suddenly and cause great loss of life and property are triggering each other today. The changing seasons and climatic conditions of climate change and global warming cause serious problems with the geographical structure and meteorology. However, in recent years, investments in risky areas, slums, and rapid population growth have begun to increase the size and impact of disasters. While disasters negatively affect human life materially and morally, they cause some problems for the economy, social life, and future plans of the country.

In the national or international context, such disasters and pandemics cause serious crises, and touristic expenditures, which are considered luxury, are directly restricted, and thus tourism movements come to a standstill. Disasters and pandemics are quite remarkable and worth researching with their results. In addition to problems such as the loss of thousands of lives, workplace, bridge, road destruction, transportation and communication disruption, dismissal of many qualified people, disruptions in education, psychological protection and panic situation, economic recession affects not only the country we live in, but the whole world, and the effects of pandemics and crises take on a global character.

The editors believe that this book will fill a huge gap in the identification, management, and control of pandemics and disasters that will always have an impact on tourism. Pandemics and natural disasters have affected people since the beginning of the historical age. Today, it affects humanity and many business lines and will continue to affect it. The book is made up of collective experiences. It provides broad coverage of issues related to pandemics and natural disasters. The book will provide academics, students, and researchers with a good and original understanding and approach. It will provide information about the problems, challenges, and future of pandemics and disasters controlling tourism.

This book is a laborious and ambitious effort by professionals and academicians from various countries. We are grateful to the authors who believed and supported us from different countries, such as India, Turkey, Malaysia, Kosovo, Pakistan, Romania and Indonesia.

Introduction

This book deals with pandemics and disasters that greatly damage the current functioning of tourism and threaten its future. The book provides theoretical and practical content for academics, researchers, students, tourism workers, business managers, tourist guides, and representatives of local and national institutions. In this context, the book focuses on the history of pandemics and disasters, tourist motivation and safety during and after pandemics and disasters, recovery of the tourism industry after pandemic and disasters, redesign, reconstruction and creation of tourism activities, crisis management and planning after pandemic and disasters, post-pandemic tourism industry recovery the routes and tourism plans, the role of governments in the management of the tourism industry, government policies, laws, and strategies after the pandemic and disasters, the role of sustainable tourism for a durable tourism after the pandemic, the current situation of the hotel industry in the new world, the marketing competitive strategies of tourism management after the pandemic, alternative tourism after the pandemic, reflections of climate change on tourism, and future trends after the pandemic. The book includes cases and practical studies that deal with the current situations, challenges, solutions, and future theories after pandemics and natural disasters.

In the first chapter, Çelikkanat Paslı addresses theoretical issues and information about the history of disasters from past to present, particularly on epidemics that have caused a pandemic and their effects on the tourism sector. In this chapter, the author presents information about disasters and pandemics from history to the present. This chapter shows the negative effects of pandemics on the tourism sector with numerical data.

In the second chapter, the authors argue critically about safety and tourist motivation in post-pandemics and disasters. According to the chapter, the tourism sector has a dynamic structure that can be affected very quickly by negative situations that may occur with the demand dimension. After the pandemic, tourism sector was affected both economically and sociopsychologically, all over the world.

In the third chapter, Sarı Gök and Aylan argue about the recovery of tourism from pandemics and disasters. The pandemic and disasters and their effects are discussed. Suggestions for the recovery of the tourism industry from pandemics and disasters are also critically presented. Studies on epidemics and natural disasters foresee that we will face natural disasters more and more, and new virus epidemics will enter into our lives within shorter periods of time and lead to more negative conclusions.

In the fourth chapter, Gündoğdu, Sharma, and Erkol Bayram focus on redesigning, reinventing, and rebuilding tourism activities in post-pandemic and disasters. This study deals with the redesigning and rediscovery of the traditional structure of the tourism sector, and the reconstruction of tourism for destinations in accordance with societies, economic and social structures, cultures and natural life. The tourism sector came to a standstill during the crisis. At this point, it was concluded that tourism should be restructured under the umbrella of sustainability and different types of tourism should be developed.

In the fifth chapter, Eren, Alnoor, and Tacu show the importance of crisis management in tourism enterprises and implementing an emergency response plan. Crisis management determines what must be done to face crises by developing plans for the crisis, implementing and evaluating the results of the plan adopted. The implementation of the emergency response plan provides important opportunities for companies in general and tourism companies in particular to achieve superior performance. Because the emergency plan positively affects the business and achieves distinguished results.

In the sixth chapter, Nazır, Kumar, Arı Ragavan, and Sharma identify the roadmap of the tourism industry post COVID-19. Embarking on the same, the current study highlights the recovery strategy of the tourism industry in the face of the COVID-19 pandemic through an extensive review of past studies. The archival analysis of existing studies has enabled researchers to set implications and recommendations for the speedy recovery of the tourism industry.

In the seventh chapter, Kumar, Nazır, and Erkol Bayram reveal the role of governments in mitigating COVID-19. Negative news about political, economic, or social status is distributed worldwide through numerous media channels. Generally, the key to success is creating a problem-solving action plan including prevention measures that combine all stakeholders' interests and activities. It is necessary to have a systematic and conceptual approach to offer suitable solutions for stakeholders—how tourism businesses react to such crises and what measures should be taken.

In the eighth chapter, Kavlak and Erkekli reveal determination of opportunities and threats, new normal policies, strategies, and regulation in the epidemic process. COVID-19 epidemic process, the negative effects of epidemics on the tourism industry can be decreased with new tourism strategies and regulations. Adaptation to new strategies usually provides a competitive advantage in a sector. At the same time, new products such as virtual travel in tourism arise with new strategies.

In the ninth chapter, Arora and Sharma focus on analyzing microfinance as a resilience strategy for poverty alleviation and promoting sustainable tourism. An attempt has also been made to create a theoretical model to guide policymakers or voluntary organizations in coming forward to initiate/support activities concerning the empowerment of the poor/deprived chunk of the rural population through microfinance.

In the tenth chapter, Singh, Agarwal, and Bhalla discuss the overwhelming reservations of customers regarding hospitality services. This study advances the understanding of the transformed management practices of the tourism and hospitality industry as a whole and the food and beverage sector in particular. The transformed management practices will help capture the altered strategies of the sector, thus addressing the customers' expectations of the new normal. The chapter presents a post-pandemic playbook with reference to the upcoming transformations and the new normal of the hospitality industry.

Yazıcı Ayyıldız and Çetin discuss the marketing strategies of hotels in the top hotel chains list of Horwath HTL's European Chains & Hotels Report 2019 in the eleventh chapter. The study shows that hotels faced major difficulties as a consequence of the COVID-19 pandemic. By adapting to the changes in terms of changing operation strategies, markets, marketing strategies, and competition strategies hotel chains have been able to cope with the changes taking place in the environment.

In the twelfth chapter, Eravcı raises awareness on market-oriented planning that cares about the consumer who is moving away from crowded living spaces due to the pandemic. A city is a market place for its residents and visitors, and as a business, it has to meet the needs of its consumers. The city tends to grow as long as it meets the needs and expectations of its consumers. The COVID-19 pandemic process has made it difficult to meet consumer needs in the city and the flow of life. Consumers and city managers have increased their concerns over the market.

In the thirteenth chapter, Shukla and Shamuraılatpam attempt to find out the role of financing by different stakeholders in the expansion of sports as an investment in human capital and how marketing can accomplish the goals as a driver of destination tourism of a particular region/state/country keeping into account different attributes of sports activities. The vital role the sports sector plays in bringing sustainable economic development in terms of building human capital, curbing crimes, revenue earnings, employability, and contribution to the GDP of the nation is taken into this research study.

In the fourteenth chapter, Bilgin, Lulaj, and Kum determine the location of financial and climate change in existing tourism plans and policies. Tourism is one of the most important sectors that contribute to the development of all countries, influencing the growth of financial and economic development. As a result of rising climate change, it is one of the most affected sectors. Although policies have been developed a lot, the issue of climate change has not been given much space and tourism strongly influences the growth of economic and financial development of countries.

In the fifteenth chapter, Korstanje argues about the opportunities and challenges of tourism in Argentina just after COVID-19. Over recent years, some voices have been alerted to the importance of resilience in the tourism industry. The global pandemic that originated in Wuhan (China) rapidly disseminated to the world, paralyzing not only the tourism industry but also global trade. COVID-19 obliges us to rethink tourism in a feudalized (atomized) world without tourists. The current travel behaviors, as well as the geographical borders, are being reformulated according to a post-pandemic situation.

In the sixteenth chapter, Bayram, Köse, and Avcı align the strategies for better tourism after disasters and pandemics. Disasters and epidemics, which are among these factors, cause sometimes short and sometimes long-term decreases in the number of tourists. Recently, it has been seen that these factors affect the attractiveness of the destination. As a result of the success shown in destination management after disasters and epidemics, the impact of material and moral damage to the region will be minimized, perhaps it will be able to gain a new attraction (ethically debatable) with tourism types such as disaster tourism or dark tourism.

In the seventeenth chapter, Boalar presents the trends and practices in domestic travel in India after COVID-19. Domestic tourism is a type which is conducted in a country. After a pandemic, many people prefer to stay in their country and this leads to a rise in domestic travel.

CHAPTER 1

History of Pandemics and Disasters and the Tourism Industry

NAZIK ÇELIKKANAT PASLI

Department of Recreational Management, Faculty of Tourism, University of Giresun, Giresun, Turkey

ABSTRACT

Disasters are events or incidents that occur naturally or caused by human effect which cause some distress or damage. A pandemic is one such disaster that can affect many people. Pandemics are epidemic diseases that spread over a wide region, sometimes even spreading around the whole globe. Such pandemics are not only crises on their own, but they also cause other crises, such as economic, political, social, and others. Pandemics fall into the "macro risks" class, which represents global and national risks, and have affected millions of people from past to present. Pandemics, such as the Black Death, cholera, Spanish flu, Asian flu, Hong Kong flu, SARS, avian flu, swine flu, MERS, Ebola, and the ongoing COVID-19 have caused damage to many people and have taken many lives around the world. Besides their effect on all sectors, these disasters have a negative impact especially on the tourism sector. Countries that are economically dependent on the tourism industry are heavily affected by the crises caused by pandemics and disasters. This chapter gives information about the history of disasters from past to present, particularly on epidemics that have caused a pandemic and their effects on the tourism sector.

1.1 INTRODUCTION

The Earth has faced many disasters throughout its lifetime of 4.5 billion years. Pandemics have particularly affected the whole world and all living

Dynamics of the Tourism Industry: Post-Pandemic and Post-Disaster Perspectives and Strategies.
Gül Erkol Bayram and Anukrati Sharma (Eds.)

beings because of the ways they spread and as detection and treatment were not always quick, there were insufficiencies, and often, people did not know what they were fighting against (Bağcı, 2020: 124). Hence, epidemics are the main disasters that have caused great loss of life and have terrified people for centuries. Such diseases are classified as endemic, epidemic, or pandemic based on their prevalence in our society. An endemic is a disease that has always been present in a population, an epidemic is one that spreads suddenly, rapidly, and unexpectedly, and a pandemic stands for a disease that is widespread in large areas like a continent or the whole world, affecting many people (Dinç, 2012: 43). Throughout the history, the development of agricultural societies has been one of the main factors in the emergence of pandemics and mass loss of life. Pandemics first emerged when large populations settled together and people were affected by diseases transmitted by animals and also when people began to coexist with the establishment of cities as finding clean water was often difficult when transitioning to a settled lifestyle, leading to increased communication between people in different regions (Özden and Özmat, 2014: 61).

In every period of human history, pandemics have been a significant factor affecting societies. Indeed, not all parts of a society are affected by a pandemic at the same level. Some segments experience the effects more, others less so. Consequently, pandemics are likely to have major effects on the subsequent evolution of the society (Türk et al., 2020: 614). Pandemics that affect large populations are fundamentally transformative due to their enormous demographic impacts and effects on production systems, as well as the capacity of societies to adapt to such external changes (Hall et al., 2020: 578).

In ancient times, human beings did not take any precautions as they did not know the cause of pandemics and regarded them as the wrath of God. Because the causes of the pandemics were not known, they could spread rapidly, and people either failed to take measures or were too late to take measures. Over time, increased knowledge and advanced technology have allowed us to have more information about pandemics. Some of the significant pandemics throughout the human history are the Black Death, cholera, Spanish flu, Asian flu, Hong Kong flu, SARS, avian flu, swine flu, MERS, Ebola, and the ongoing COVID-19.

1.2 PANDEMICS AND THEIR EFFECTS ON TOURISM

There have been major pandemics of plagues throughout history, each time they caused great loss of life. These plagues are often called the "Black

Death" or the "Black Plague" because of the typical characteristic of black-colored swelling. The first recorded plague occurred in Egypt in 541 AD. This pandemic had many economic, social, and religious effects and was named the "plague of Justinian" or the "Justinian plague" (541–542 AD). It was named after the emperor of the Byzantine empire because it weakened the empire. It is estimated that about 50–60% of the population of North Africa, Europe, and Central and South Asia lost their lives in this pandemic (Parıldar, 2020: 22). Another pandemic in 1348 was perhaps the strongest of all plagues, which wiped out almost a third of Europe's population. It is often considered that the plague first emerged in China, advanced around Lake Baikal and the Lower Volga region, and passed to Feodosia. In 1345, the Mongolian armies who besieged the Genoese colony in Crimea hurled plagued dead bodies into the city with catapults. At the time, Feodosia was frequented by European traders. When 12 Genoese ships that traded with Feodosia brought their goods to the port of Messina, Sicily, the disease spread to the European continent. The most important point about the Black Death is its effects on medieval Europe. After the plagues, the population of Florence fell from 120,000 in 1330 to 37,000 in 1427. The plague deeply affected Europe socially as well as deteriorated peoples' trust in the church. The European population was around 76 million by 1340 and 50 million by 1450. Then, between 1720 and 1722, the Great Plague of Marseille occurred, which began in France, killing 75–200 million people. The next pandemic is estimated to have spread from China in 1855. It is believed that 12.5 million Indians died between 1898 and 1918. After the 1950s, plague pandemics were slowed down using preventive measures and antibiotics, but there were small epidemics (Porter, 2009: 7; Genç, 2011: 125–131; Nikiforuk, 2020: 67–91; Parıldar, 2020: 22).

With the end of the 19th century, the Black Plague had lost its dominance on the European lands and left its place to cholera. Cholera is an infectious disease that presents with a severe intestinal infection. It could turn into a major pandemic in countries that do not have adequate sanitation equipment and access to adequate clean drinking water, where sewer systems are not fully established, or where the infrastructure is damaged due to war and destruction. The year 1817 was a turning point for cholera pandemics in the world. Until then, the disease was rare, affecting only small regions, but now it was in the form of large epidemics. The world has seen seven major pandemics since then, namely, during the years 1826–1837, 1840–1849, 1856, 1863–1865, 1879–1884, 1891–1896, 1914–1918 (Uyanık and Aksoy, 2016: 230).

In the 20th century, the tourism industry was negatively affected by many pandemics besides the blow it received from the World Wars. Especially, as railway transportation became more popular, increasing travel further accelerated the spread of diseases (Kučerová and Bělousovová, 2021: 16). Of the three flu pandemics that emerged in the 20th century, the most severe pandemic has been the "Spanish flu," which is estimated to have caused 20–50 million deaths between 1918 and 1919. The other two flu pandemics were the "Asian flu" in 1957–1958 and the "Hong Kong flu" in 1968. These two flu pandemics are estimated to have caused 1–4 million deaths each (WHO, 2022a).

The Spanish flu began in the spring and summer of 1918 and occurred in three waves. These waves spread around nearly the whole globe in less than a year. Indeed, not all countries experienced the flu in all three waves. Australia, for example, experienced a single, longer wave, due to the partial success of a maritime quarantine that delayed the pandemic until early 1919 (Johnson and Mueller, 2002: 107). Although there is no exact death toll, the pandemic known as the "Spanish flu" is estimated to have caused 50–100 million deaths around the world (Ashton, 2020: 197). The flu was not spread by only military personnel and migrant workers. Other factors were also involved, for example, there were traditional holidays in some Spanish towns and villages during this period; people celebrated these holidays with parties with high participation. It is believed that almost everyone who participated in these activities caught the disease within a few days (Trilla et al., 2008: 670). Europe and the United States have made efforts in severely understaffed civilian hospitals, partly because doctors and nurses were involved in the war and due to the large number of people who contracted the disease. As a result, the Spanish flu demonstrates that healthcare facilities have failed to contain the spread of the pandemic and to provide efficient care (Aassve et al., 2020: 843). This flu pandemic caused the death of millions within a year and a half as it emerged suddenly and turned into a nightmare for people. The H1N1 virus that caused this flu affected an unexpected target population which included healthy youth and adults, rather than the weak, the elderly, or the children. Encountered during World War I, this flu is called the "Spanish flu" because it was first identified in Spain (Uyanık and Aksoy, 2016: 234). Virologists and epidemiologists around the world later determined that the Spanish flu in fact did not emerge in Spain, approximately 90 years after the pandemic began (Trilla et al., 2008: 668). World War I, which started in 1914 and ended in 1918, brought tourism travel to a standstill during this period. Coinciding with the end of the war, the Spanish flu quickly affected many countries, therefore not allowing the revival of postwar tourism.

Due to the increasing population, growing cities, and the destruction of forests, human life has been more intertwined with wildlife, and more pandemics have arisen in the 21st century. With the development of travel networks, pandemics can now affect the whole world much faster than before. Quarantine measures, social distancing, travel restrictions, and postponing or canceling events can seriously affect tourism. In short, travel and tourism have played a major role in the spread of pandemics and they have suffered major harm due to the restrictions and the fear of tourists (Zeydan and Gürbüz, 2020: 130; Bayram et al., 2021: 39).

Severe acute respiratory syndrome, also known as SARS, was first presented in China in 2002–2003 and had spread over more than 20 countries (Bağcı, 2020: 129). With most of its cases in China and Hong Kong, as well as some clusters in Taiwan and Canada, SARS was identified as a pandemic by the World Health Organization (Gössling et al., 2021: 4). The World Health Organization reported a summary of SARS cases from November 1, 2002 to July 31, 2003, announcing the death of 774 people out of 8096 infected cases in 25 countries/regions, particularly in China, Hong Kong, Taiwan, Canada, and Singapore (WHO, 2015). SARS was first identified in southern China. The world became aware of the disease in March 2003, when a tourist in Hong Kong infected several people staying at the same hotel. Seven of the infected guests returned to their homes in Canada, Vietnam, and Singapore, where they spread the disease. During the pandemic, the rest of Asia and even countries that were largely or completely free of the disease experienced at least a 70% decline in incoming tourists (McKercher and Chon, 2004: 716). In the early stages of the pandemic, tourists were primarily affected and they became the spreaders of the disease. Due to the ease of transmission, the speed of spread, and media influence, the pandemic soon led to international concerns. On March 15, 2003, a series of travel recommendations were published by the World Health Organization, including postponing nonessential travel to regions that are affected by SARS. The objective was to limit the spread of infection by international travel. Between April 1 and 21, tourist arrivals in East Asia decreased by 41% compared with the same period in 2002. China, Hong Kong, Vietnam, and Singapore suffered particularly severe damage. During the pandemic, Asia and the Pacific lost 12 million visitors compared with the previous year. Hotel occupancy rates in Beijing dropped by as much as 10%. Due to the decrease in overseas and domestic tourist arrivals to Beijing during the first 5 months of 2003, the country suffered an economic loss of USD 1.3 billion (Wilder-Smith, 2006: 55). Hong Kong experienced a 7.2% decrease in tourist arrivals compared with the previous year (Otoo and Kim, 2018: 347). In Singapore, tourism hit its lowest level in 2003, again mainly due to the fear of SARS (Lee and Chen, 2011: 1422). A Russian airline, which flew twice a week to Beijing stopped its flights due to the fear of the deadly SARS virus. An airline company in Turkey rejected boarding passengers who had a body temperature of 38 or above among those arriving from Singapore, Shanghai, and Beijing on its Hong Kong, Bangkok, and connecting flights. Besides, Turkey advised its citizens not to go to China, Hong Kong, Taiwan, Singapore, Vietnam, Philippines, Indonesia, or Thailand unless for urgent reasons. A national airline in Switzerland reduced the number of its flights to the Middle East and Asian regions due to the war and SARS, stopping flights to some cities for a while. The company canceled its flights to cities, such as Beijing, Hong Kong, Johannesburg, Boston, Washington, and Cairo, also was reducing flights to Zurich, Athens, Frankfurt, Berlin, Paris, Rome, Belgrade, London, Vienna, Brussels, and Istanbul. In the first days of SARS, the state airline of Indonesia announced a 20% decrease in passengers. In Shanghai, China, where SARS originated, four major accommodation businesses were closed for 3 months (Küçükaltan et al., 2015: 35–36). Some countries, particularly those that were directly affected by the pandemic, suffered a decrease in tourism. Other countries, where there was no pandemic, observed a slower increase in tourism. In 2003, the number of tourists increased by 5% and tourism revenues increased by 10% compared with the previous period in Turkey. Even though the SARS pandemic emerged in Far Eastern countries, Turkey still experienced a lower increase in tourism values during this period (Göçen et al., 2011: 508).

Avian flu, also called bird flu, is a deadly strain of influenza transmitted through poultry, caused by the H5N1 virus (Uyanık and Aksoy, 2016: 237). The first case of bird flu was reported in Hong Kong in 1997. Six years later, in 2003, the first new case was detected in China. Research showed that the disease was not transmitted between humans; however, some countries still issued travel warnings to their citizens to avoid all unnecessary travel to affected regions. Visitors to Asian countries were informed to avoid contact with and consumption of poultry (Lee and Chen, 2011: 1421).

The pandemic had significant economic effects on the agricultural sectors of the affected countries. With the first new cases in December 2003, poultry populations declined drastically in some parts of Asia and Europe (Shaluf, 2007: 694). This pandemic affected other poultry farms as well, especially duck farms, causing a great financial burden. Poultry farmers and producers in other regions were also hurt by the overall increased demand for ducks and other poultry products, as prices fell sharply and there was panic among consumers. The pandemic also affected stakeholders in some subsectors like the feed and pharmaceutical industries. For example, in Kerala, India, the government invested a significant amount to control the disease. The transportation, accommodation, and entertainment sectors of Kerala were adversely affected, even including boat trips on still water (Govindaraj et al., 2017: 2). In 2003, when bird flu emerged, the number of visitors decreased by 4.29% in Germany, 9.98% in France, 17.74% in Luxembourg, 19.61% in New Zealand, 22.86% in the Czech Republic, 25.99% in Finland, 26.41% in Portugal, 28.19% in Japan, 32.20% in Poland, and 64.44% in Denmark compared with the previous year. In 2006, when it emerged in Turkey, the number of tourists decreased by 6% and tourism revenues decreased by 5% compared with the previous year. Considering other countries, in 2006, the number of tourists decreased by 0.60% in Russia, 4.49% in England, 6.18% in France, 9.54% in Iran, 11.34% in Germany, 20.46% in the Netherlands, and 27.38% in Bulgaria compared with the previous year (Göçen et al., 2011: 506–508).

SARS had a bigger impact on international tourist arrivals than the bird flu because the bird flu was transmitted from animals to humans. So, it was not contagious among humans. If the bird flu could be easily transmitted between people, like SARS, it would have caused bigger concerns among people, and travel restrictions would have been even more prominent (McAleer et al., 2010: 100–102). Table 1.1 depicts how SARS and bird flu affected countries.

TABLE 1.1 Human Cases of SARS and Avian Flu by Country.

	SARS			Avian Flu		
Country	**Infection**	**Deaths**	**Death rate (%)**	**Infection**	**Deaths**	**Death rate (%)**
Cambodia	–	–	–	7	7	100.00
China	5327	349	6.55	22	13	59.09
Hong Kong	1755	299	17.04	–	–	–
Indonesia	2	0	0	101	81	80.20
Korea	3	0	0	–	–	–
Malaysia	5	2	40.00	–	–	–
Singapore	238	33	13.87	–	–	–
Taiwan	346	37	10.69	–	–	–
Thailand	9	2	22.22	25	17	68.00
Turkey	–	–	–	12	4	33.33
Vietnam	63	5	7.94	95	42	44.21
Our sample	7748	727	–	262	164	–
Asia	7783	729	–	279	176	–
World	8096	8096		319	192	–
Time period	Jan 2003–Dec 2003			Jan 2004–July 2007		

Source: Adapted from McAleer et al. (2010: 102).

Swine flu was first seen in North America (Mexico) in April 2009, later spreading rapidly around the world. As the World Health Organization declared it a pandemic in June 2009, swine flu was confirmed in 74 countries and regions, with reported infections (WHO, 2022b). The tourism industry in Mexico was heavily affected by swine flu. During this period, many travel agencies abroad canceled their flights and vacation deals to Mexico. Countries, such as Argentina and Cuba also suspended all flights from Mexico. In short, many trips to and from Mexico were canceled due to the pandemic, and other businesses related to the tourism industry were closed along with numerous public spaces and attractions (Garg, 2013: 51). This stagnation in tourism was also evident in air travel. Hotel occupancy rates in Mexico fell about 9–12% during May and early June. In Cancún, hotel occupancy rates were near 75% in April, before the swine flu outbreak, which dropped to around 20% in May and early June. This forced 22 hotels in Cancún to suspend their operations (LADB Staff, 2009). In Argentina, the media highlighted that the main cause of infection was tourism and mass travel, urging experts to advise against any travel (Maximiliano, 2010: 171). Even a small country like Brunei lost about 30,000 (15%) tourists and 15 million USD in the first 12 months after swine flu. To sum up, the number of tourists decreased significantly due to both swine flu and the global financial crisis that coincided with the pandemic (Haque and Haque, 2018: 99).

The MERS outbreak was first reported in Saudi Arabia in September 2012, but research showed that the first cases emerged in Jordan in April 2012. The largest known MERS outbreak outside the Arabian peninsula occurred in the Republic of Korea in 2015. Here, it is noteworthy that one person from outside the country caused the pandemic (Parıldar, 2020: 20). Since 2012, the disease was identified in 2578 cases in 27 countries, causing 888 deaths in total (WHO, 2021). During the MERS outbreak in Korea in 2015, the number of noncitizen visitors decreased significantly between June and September. The pandemic caused a decrease of 2.1 million (16%) in foreign tourists visiting Korea, reducing tourism revenues by 2.6 billion USD. During this period, the greatest decrease occurred in July 2015 (Joo et al., 2019: 3).

The Ebola virus outbreak has been dominant in Central and West Africa, particularly in the Democratic Republic of the Congo (Maphanga and Henama, 2019: 2). The first Ebola pandemic broke out in 1976 in the two poorest regions of Africa: Sudan and Zaire. Although this was not the first Ebola pandemic, it was the first to affect both Europeans and

Africans (Nikiforuk, 2020: 262). The 2014–2016 outbreak in West Africa is considered the largest and most complex Ebola outbreak since the virus was discovered in 1976. During this period, the pandemic spread between countries, from Guinea to Sierra Leone and Liberia (WHO, 2022c). Since the outbreak of the Ebola virus in Guinea, Liberia, and Sierra Leone in May 2014, the tourism and hospitality industry in Sierra Leone were adversely affected. Most businesses were closed, hotel reservations and flights were canceled, and restrictions were imposed on travelers. The exchange rate and revenues dropped and many businesses related to tourism, such as fishing and vegetable growing were negatively affected (Kongoley-MIH, 2015: 543). As flights from the UK and the Netherlands decreased by 50%, local tourism stakeholders (hotels in particular) began to implement cost-cutting strategies to reduce the impact of the crisis on their businesses. For example, a hotelkeeper reported not hiring an operations manager and rotating all seasonal staff. The crisis also caused the delay or cancellation of new hotel constructions (Novelli et al., 2018: 83). Therefore, the decline in tourism due to the fear of contracting Ebola negatively affected the economy of Africa (Maphanga and Henama, 2019: 7). This pandemic had a significant impact on the world and particularly on West Africa. Overall, 28,616 Ebola cases and 11,310 deaths were reported in Guinea, Liberia, and Sierra Leone. When the pandemic spread outside these three countries, 36 more cases and 15 more deaths were reported (CDC, 2019).

COVID-19 was first identified on January 13, 2020, in a group of patients who developed respiratory symptoms (fever, cough, and shortness of breath) in Wuhan, China, in late December. The disease was initially detected in people who were in the seafood and animal market in this region. Later, the disease spread to other cities in Hubei and then to the other states of the People's Republic of China, and soon after to other countries (Sağlık Bakanlığı, 2022).

Due to the COVID-19 pandemic, there have been many measures taken to prevent mobility, which negatively affected the service sector. Tourism businesses have particularly stood out as they had to completely halt their activities. As the tourism sector requires the physical participation of people, the business has been uncertain. With the pandemic, many tourism businesses lost their revenues and many employees lost their jobs (Atay, 2020: 168).

As of January 31, 2022, there have been 373,229,380 confirmed COVID-19 cases and 5,658,702 deaths reported to the World Health Organization globally. As of the same date, 9,901,135,520 doses of vaccines have been administered (WHO, 2022d).

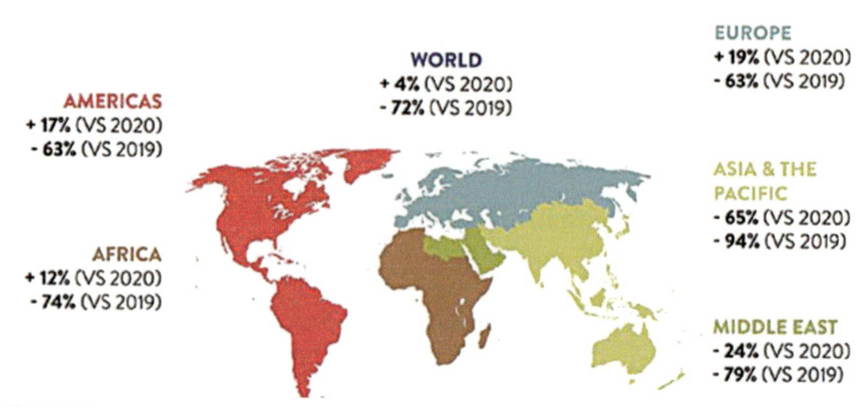

FIGURE 1.1 International tourist arrivals by region.

Source: Adapted from UNWTO (2022).

With increased demand, rapid progress in vaccines, and facilitated entry restrictions in many destinations, there were 15 million more tourist visits in 2021 than in 2020. These visits increased the volume of international tourism by 4% in 2021, although international arrivals were still below by 72% in 2019. International tourism began to show a moderate recovery in the second half of 2021, though compared with that before the pandemic, international arrivals still decreased by 62% in the third and fourth quarters. The full impact of the Omicron variant and the possible fluctuations in the number of COVID-19 cases are yet to be seen. For the tourism sector, the pace of recovery remains slow and uneven across the world, as travel restrictions, vaccination rates, and passenger safety all vary between countries. Europe (19%) and America (17%) recorded the highest mobility in international tourism in 2021 compared with 2020, though still below the previous years by 63%. Regarding the other regions, the Caribbean region experienced a 63% increase compared with 2020, although still at a 37% loss compared with 2019. Southern Mediterranean Europe (57%) and Central America (54%) experienced a significant recovery, despite being below by 54% and 56%, respectively in 2019. North America (17%) and Middle East Europe (18%) also exceeded the numbers in 2020. Africa, on the other hand, saw a 12% increase in arrivals compared with 2020, but still below by 74% in 2019 (UNWTO, 2022). In short, the mobility of international tourism has improved in 2021 compared with 2020 in many regions of the world, while this improvement is still very low compared with the years before the pandemic.

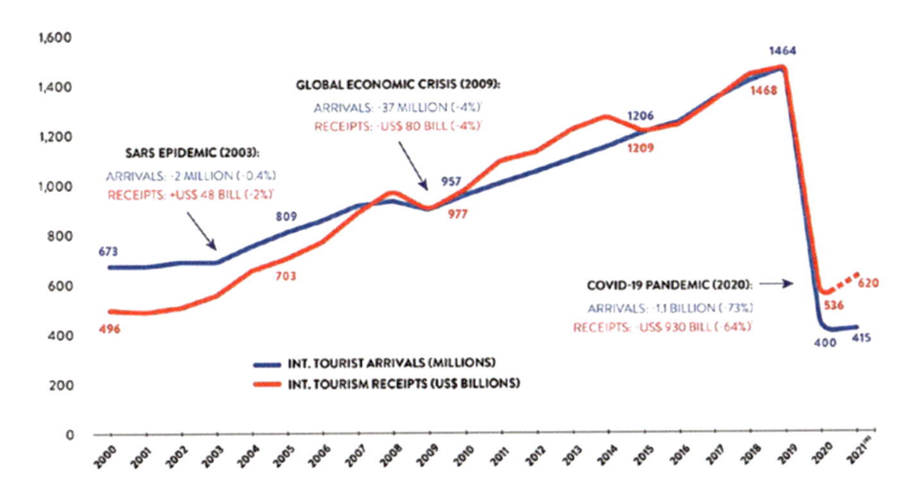

FIGURE 1.2 International tourist arrivals and tourism receipts 2000–2021.

Source: Adapted from UNWTO (2022).

In 2021, the economic contribution of tourism (measured directly as gross domestic product) was estimated at USD 1.9 trillion. This is higher than that of 2020 by USD 1.6 trillion, although still lower than the year before the pandemic. Export revenues from international tourism could exceed USD 700 billion in 2021, which would be a small improvement compared with 2020 due to higher expenditures per trip, but it is less than half of the USD 1.7 trillion in 2019. Spending per capita was at USD 1300 in 2020 and is expected to reach nearly USD 1500 in 2021. This is due to longer stays and higher transportation and accommodation prices (UNWTO, 2022) Also, considering the canceled pilgrimage visits in 2020 by Saudi Arabia due to COVID-19, there have been great losses in religious tourism as well, accounting for around USD 8.5 billion (Mubarak and Zin, 2020).

In short, emerging in late 2019 and affecting the whole world, COVID-19 had some negative impact on the tourism sector. Countries closed their borders and banned entrance and exit for tourists, implementing travel restrictions and curfews as a seasonal measure, further preventing travel. Considering the circumstances, tourism businesses had to make some decisions. Some businesses temporarily ceased activities, while others took the opportunity to carry out renewal processes that they had been waiting for; they were unable to perform these processes before because their businesses were active for all 12 months. Some contributed to healthcare services by offering their businesses to the use of healthcare workers within social responsibility projects (Paslı and Vapur, 2020: 8).

To sum up, the spread of viruses causes diseases to easily infect others and spread from one region to another. In such circumstances, people who are concerned about their health will be reluctant toward tourism activities. With their global structure, tourism activities also cause viruses to spread quickly and easily between societies. This makes it necessary to take certain measures on a global scale. One such measure is to introduce national/international or regional travel restrictions. Thus, tourism has a structure that both affects and is affected by pandemics, as seen in Figure 1.3 (Kiper et al., 2020: 529).

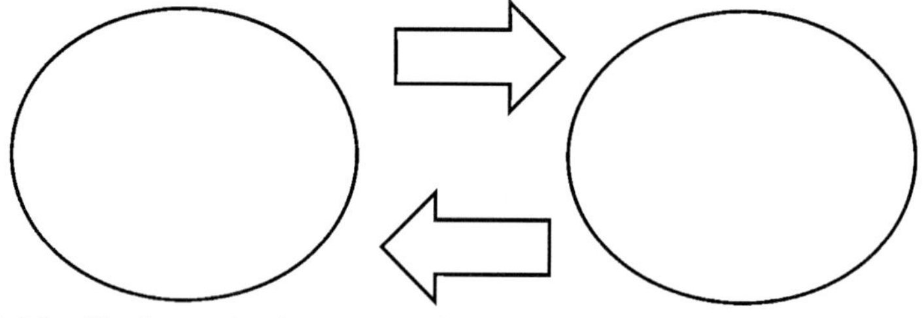

FIGURE 1.3 The interaction between pandemics and tourism activities.

Source: Adapted from Kiper et al. (2020: 529).

One of the negative consequences of emerging health crises for the tourism sector is the decreased demand for destinations (countries where pandemics are most common). Due to the pandemics so far, the image of countries where such diseases have emerged has been negatively affected (Çeti and Ünlüönen, 2019: 122–123). Factors like pandemics may cause a region to be quarantined, not only preventing tourists from coming in, but also restricting citizens from participating in tourist activities (Çakmak, 2019: 72).

1.3 CONCLUSIONS

Pandemics have a long history, dating back to the ancient times. Human beings have always struggled with pandemic diseases and millions of people have lost their lives because of them. In the past, people defined such diseases as "the wrath of God, the revenge of nature, etc." However, with advanced medicine, science, and technology, they learned more about these diseases and even found solutions to prevent or cure most of them. Throughout history, there have been some significant pandemics, such as the Black Plague, cholera, Spanish flu, Asian flu, Hong Kong flu, SARS, avian flu, swine flu, MERS, Ebola, and COVID-19. These diseases affected not only the infected individuals, but also whole societies, economies, many industries, countries, and even future generations.

With its high multiplier effect, tourism impacts over 135 sectors, contributing to the payment balance by creating a foreign currency inflow and sparking many other leading industries, such as transportation, trade, construction, health, and finance. Apart from its economic effects, tourism also has social and cultural effects (Çetin and Göktepe, 2020: 90). Humans are social creatures. Apart from basic needs, such as nutrition and shelter, we have other needs, such as education, entertainment, artistic activities, and tourist trips. Pandemics negatively affect social life. Due to measures, such as travel restrictions, closure of schools, and postponement of sports competitions and artistic activities (theaters, concerts, etc.) during such periods, daily life becomes different, disrupting the functioning of social activities. Infectious diseases harm people's lives and livelihoods, just like wars and financial crises. Hence, preventing pandemic diseases should be considered not only as a health issue but also as the basic principle of national and global security (Özkoçak et al., 2020: 1189–1190).

Considering historical processes, the timing of a pandemic seems uncertain, although they have always been experienced and will be experienced in the future. Famous writer George Santayana says, "Those who cannot remember the past are condemned to repeat it." Humanity must take precautions and be prepared for pandemics. The whole world needs to unite on this issue. In the globalizing world, no one can escape from a pandemic by simply closing their doors, which they had to learn in the most difficult way by experiencing various pandemics throughout history, and even today, with the ongoing COVID-19. There should be pandemic scenarios for the coming years and no country should be allowed to be caught unprepared by one. Measures should be taken to prevent pandemics and sanctions should be applied. When taking these measures, we should learn our lesson from history, eliminate the necessary deficiencies, and reorganize the steps to be implemented accordingly.

This chapter shows the negative effects of pandemics on the tourism sector with numerical data. The ongoing COVID-19 pandemic, the decreasing number of tourists, restricted travel, and quarantines have adversely affected both people who want to travel and the stakeholders of the tourism sector.

To prevent the social, economic, and medical damage that will arise from similar pandemics in the coming years, the necessary measures should be taken to protect tourism stakeholders.

KEYWORDS

- **pandemics**
- **history of pandemics**
- **pandemics and their effects on tourism**

REFERENCES

Aassve, A.; Alfani, G.; Gandolfi, F.; Le Moglie, M. Epidemics and Trust: The Case of the Spanish Flu. *Health Econ.* **2020,** *30* (4), 840–857.

Ashton, J. COVID-19 and the 'Spanish' Flu. *J. R. Soc. Med.* **2020,** *113* (5), 197–198.

Atay, L. COVID-19 Salgını ve Turizme Etkileri. *Seyahat ve Otel İşletmeciliği Dergisi* **2020,** *17* (1), 168–172.

Bağcı, B. *Turizmde Kriz*; Cinius Yayınları: İstanbul, 2020.

Bayram, M.; Arıcı, S.; Bayram, Ü. *COVID-19 ve Turizm: Ülkelerin Kriz Yönetim Stratejilerinin Karşılaştırmalı Analizi*; Gazi Kitabevi: Ankara, 2021.

CDC. https://www.cdc.gov/vhf/ebola/history/2014–2016-outbreak/index.html (accessed Jan 7, 2022).

Çakmak, T. F. *Turizm Endüstrisinde Bütüncül Yaklaşımla Kriz Yönetimi ve Örnek Olaylar*; Detay Yayıncılık: Ankara, 2019.

Çeti, B.; Ünlüönen, K. Salgın Hastalıklar Sebebiyle Oluşan Krizlerin Turizm Sektörü Üzerindeki Etkisinin Değerlendirilmesi. *AHBVÜ Turizm Fakültesi Dergisi* **2019,** *22* (2), 109–128.

Çetin, G.; Göktepe, S. COVID-19 Pandemisinin Turizm Endüstrisi Üzerindeki Etkileri. In *COVID-19 Pandemisinin Ekonomik, Toplumsal ve Siyasal Etkileri*; Demirbaş, D., Bozkurt, V., Yorğun, S., Eds.; İstanbul Üniversitesi Yayınları: İstanbul, 2020, pp. 87–97.

Dinç, G. Bulaşıcı ve Salgın Hastalıklar Tarihine Genel Bir Bakış, *Yeni Tıp Araştırmaları* **2012,** *18*, 43–72.

Garg, A. A Study of Tourist Perception Towards Travel Risk Factors in Tourist Decision Making. *Asian J. Tour. Hosp. Res.* **2013,** *7* (1), 47–57.

Genç, Ö. *Kara Ölüm: 1348 Veba Salgını ve Ortaçağ Avrupa'sına Etkileri. Tarih Okulu Dergisi* **2011,** *X*, 123–150.

Govindaraj, G.; Sridevi, R.; Nandakumar, S. N.; Vineet, R.; Rajeev, P.; Binu, M. K.; Balamurugan, V.; Rahman, H. Economic Impacts of Avian Influenza Outbreaks in Kerala. *Transboundary Emerg. Dis.* **2017,** *65* (2), 1–12.

Göçen, S.; Yirik, Ş.; Yılmaz, Y. Türkiye'de Krizler ve Krizlerin Turizm Sektörüne Etkileri. *Süleyman Demirel Üniversitesi İktisadi ve İdari Bilimler Fakültesi Dergisi* **2011,** *16* (2), 493–509.
Gössling, S.; Scott, D.; Hall, C. M. Pandemics, Tourism and Global Change: A Rapid Assessment of COVID-19. *J. Sustain. Tour.* **2021,** *29* (1), 1–20.
Hall, C. M.; Scott, D.; Gössling, S. Pandemics, Transformations and Tourism: Be Careful What You Wish For. *Tour. Geogr.* **2020,** *22* (3), 577–598.
Haque, T. H.; Haque, M. O. The Swine Flu and Its Impact on Tourism in Brunei. *J. Hosp. Tour. Manage.* **2018,** *36*, 92–101.
Johnson, N. P.; Mueller, J. Updating the Accounts: Global Mortality of the 1918–1920 "Spanish" Influenza Pandemic. *Bull. Hist. Med.* **2002,** *76*, 105–115.
Joo, H.; Maskery, B. A.; Berro, A. D.; Rotz, L. D.; Lee, Y.-K.; Brown, C. M. Economic Impact of the 2015 MERS Outbreak on the Republic of Korea's Tourism- Related Industries. *Health Secur.* **2019,** *17* (2), 1–9.
Kiper, V. O.; Saraç, Ö.; Çolak, O.; Batman, O. COVID-19 Salgınıyla Oluşan Krizlerin Turizm Faaliyetleri Üzerindeki Etkilerinin Turizm Akademisyenleri Tarafından Değerlendirilmesi. *Balıkesir Üniversitesi Sosyal Bilimler Enstitüsü Dergisi* **2020,** *23* (43), 527–551.
Kongoley-MIH, P. S. The Impact of Ebola on the Tourism and Hospitality Industry in Sierra Leone. *Int. J. Sci. Res. Pub.* **2015,** *5* (12), 542–550.
Kučerová, J.; Bělousovová, N. Pandemics and Their Impact on Tourism. Scientia Iuventa, 2021. Book of Extended Abstracts from the International Scientific, 15.
Küçükaltan, D.; Aydın Tükeltürk, Ş.; Çiftçi, G. *Otel İşletmelerinde Kriz Yönetimi*; Detay Yayıncılık: Ankara, 2015.
LADB Staff. Negative Impact on Mexican Tourism Continues From April Outbreak of Swine Flu, LADB, Latin American Data Base News & Educational Services, 2009.
Lee, C.-C.; Chen, C.-J. The Reaction of Elderly Asian Tourists to Avian Influenza and SARS. *Tour. Manage.* **2011,** *32*, 1421–1422.
Maphanga, P. M.; Henama, U. S. The Tourism Impact of Ebola in Africa: Lessons on Crisis Management. *Afr. J. Hosp. Tour. Leisure* **2019,** *8* (3), 1–13.
Maximiliano, K. Role of Mass-Media in Swine Flu Outbreak in Buenos Aires. *Anatolia: Int. J. Tour. Hosp. Res.* **2010,** *21* (1), 169–178.
McAleer, M.; Huang, B.-W.; Kuo, H.-I.; Chen, C.-C.; Chang, C.-L. An Econometric Analysis of SARS and Avian Flu on International Tourist Arrivals to Asia. *Environ. Model. Softw* **2010,** *25*, 100–106.
McKercher, B.; Chon, K. The Over-Reaction to SARS and the Collapse of Asian Tourism. *Ann. Tour. Res.* **2004,** *31* (3), 716–719.
Mubarak, N.; Zin, C. S. Religious Tourism and Mass Religious Gatherings—The Potential Link in the Spread of COVID-19. Current Perspective and Future Implications. *Travel Med. Infect. Dis.* **2020,** *36*, 101786.
Nikiforuk, A. *Mahşerin Dördüncü Atlısı: Salgın ve Bulaşıcı Hastalıklar Tarihi*. 9; Baskı, İletişim Yayınları: İstanbul, 2020.
Novelli, M.; Burgess, L. G.; Jones, A.; Ritchie, B. W. 'No Ebola… Still Doomed'—The Ebola-Induced Tourism Crisis. *Ann. Tour. Res.* **2018,** *70*, 76–87.
Otoo, F. E.; Kim, S. Is There Stability Underneath Health Risk Resilience in Hong Kong Inbound Tourism? *Asia Pacific J. Tour. Res.* **2018,** *23* (4), 344–358.
Özden, K.; Özmat, M. Salgın ve Kent: 1347 Veba Salgınının Avrupa'da Sosyal, Politik ve Ekonomik Sonuçları. *İdealkent* **2014,** *12*, 60–87.

Özkoçak, V.; Koç, F.; Gültekin T. Pandemilere Antropolojik Bakış: Koronavirüs (COVID-19) Örneği. *Turk. Stud.* **2020,** *15* (2), 1183–1195.

Parıldar, H. Tarihte Bulaşıcı Hastalık Salgınları. *Tepecik Eğit. ve Araşt. Hast. Dergisi* **2020,** *30* (Ek sayı), 19–26.

Paslı, M. M.; Vapur, M. Risk Kavramı. In *Turizmde Risk Yönetimi*; Tuna, M., Ed.; Detay Yayıncılık: Ankara, 2020, pp. 3–33.

Porter, S. *The Great Plague*; Amberley Publishing: Great Britain, 2009.

Sağlık Bakanlığı. https://covid19.saglik.gov.tr/TR-66300/COVID-19-nedir-.html (accessed Jan 27, 2022).

Shaluf, I. M. An Overview on Disasters. *Disaster Prev. Manag. Int. J.* **2007,** *16* (5), 687–703.

Trilla, A.; Trilla, G.; Daer, C. The 1918 "Spanish Flu" in Spain. *Clin. Infect. Dis.* **2008,** *47* (5), 668–673.

Türk, A.; Ak Bingül, B.; Ak, R. Tarihsel Süreçte Yaşanan Pandemilerin Ekonomik ve Sosyal Etkileri. *Gaziantep Univ. J. Soc. Sci.*, Special Issue **2020,** 612–632.

UNWTO. https://www.unwto.org/impact-assessment-of-the-COVID-19-outbreak-on-international-tourism (accessed Jan 27, 2022).

Uyanık, V.; Aksoy, Y. *Tarihteki Büyük Felaketler*; Siyah Beyaz Yayınları: İstanbul, 2016.

Zeydan, İ.; Gürbüz, A. 21. Yüzyıldaki Pandemiler ve Turizm Sektörüne Etkileri, In *Econder, 2020 3rd. International Economics, Business and Social Sciences Congress*; 2020; pp 126–132.

WHO. http://www.emro.who.int/health-topics/mers-cov/mers-outbreaks.html (accessed Dec 29. 2021).

WHO. https://www.who.int/publications/m/item/summary-of-probable-sars-cases-with-onset-of-illness-from-1-november-2002-to-31-july-2003 (accessed Jan 05. 2022a).

WHO. https://www.euro.who.int/en/health-topics/communicablediseases/influenza/pandemic-influenza/past-pandemics (accessed Jan 07. 2022b).

WHO. https://www.who.int/emergencies/situations/influenza-a- (h1n1)-outbreak (accessed Jan 23. 2022c).

WHO. https://www.who.int/health-topics/ebola/#tab=tab_1 (accessed Jan 25. 2022d).

WHO. https://covid19.who.int/ (accessed Jan 31. 2022e).

Wilder-Smith, A. Tourism and SARS. In *Tourism in Turbulent Times: Towards Safe Experiences for Visitors*; Wilks, J., Pendergast, D., Leggat, P., Eds.; Elsevier Ltd., 2006, pp 53–61.

CHAPTER 2

Safety Motivation and Tourism Post-Pandemic and Post-Disasters

MELIKE SAK[1], MUTHMAINNAH[2], ANDI ASRIFAN[3], and ABDUL WAFI[4]

[1]*Instution of Social Sciences, Travel Management and Tourism Guiding, Selcuk University, Konya, Turkey*

[2]*Indonesian Language Department, Teacher Training and Education Faculty, Universitas Al Asyariah Mandar, Indonesia*

[3]*English Language Department, Universitas Muhammadiyah Sidenreng Rappang, Indonesia*

[4]*Institut Agama Islam Negeri (IAIN) Madura, Tarbiyah Faculty, Indonesia*

ABSTRACT

Tourism means all the activities that people participate in for various purposes, such as seeing cultural and historical destinations, trying new food and experiencing new tastes, traveling to new places, and resting (Emekli, 2006). People primarily consider whether it is safe or not when deciding on the trips they will take to rest and see new places. In the decision of tourists to travel for the purposes of resting, having fun, getting to know different cultures, hygiene, health, and safety are among the most important factors that determine the holiday motivation of tourists. In this context, we can state that the tourism sector has a dynamic structure that can be affected very quickly by the negative situations that may occur with the demand dimension.

Dynamics of the Tourism Industry: Post-Pandemic and Post-Disaster Perspectives and Strategies.
Gül Erkol Bayram and Anukrati Sharma (Eds.)

The pandemic, which emerged in the city of Wuhan, People's Republic of China, on December 31, 2019 was named "coronavirus" or COVID-19 by the World Health Organization (WHO), affected the whole world. The fact that many deaths have occurred in the world, and diseases of the respiratory tract has led to deaths shows that the epidemic had reached serious dimensions. The negative situation it has created in terms of health has caused many damages in terms of economic and social aspects (Kaygısız, 2021). COVID-19 epidemic has affected the whole world and has completely disrupted the communication of people and their interaction with each other. COVID-19 epidemic hit the social and economic assets of countries along with the tourism sector, which is one of the most affected. The COVID-19 epidemic has seriously affected the tourism sector, causing tourism activities all over the world to stop completely with the prohibition of international travel. The coronavirus spreads faster in crowded environments and in areas where people interact. By restricting activities in museums and tourist destinations, it ruins the business in places where crowded and social distance areas cannot be protected. Hotel rooms are left empty as a result of reservation cancellations in accommodation businesses. In addition, it has resulted in the closure of food and beverage businesses (cafes, restaurants, restaurants) for hygiene reasons, cancellation of flight and bus tickets and transportation services. Measures have been taken to protect human health by visa cancellations, countries closing their borders, curfews, imposing restriction of domestic and international travel, and practicing social distance rules (Kıvılcım, 2020).

Even if the bans and restrictions on the COVID-19 epidemic were gradually lifted over time and countries entered the normalization process, people began to worry about entering crowded environments and traveling again. The anxiety and worry caused by the epidemic have affected the travel behavior and security motivations of tourists. In the face of this situation, to increase the demand for the tourism sector, reviving and generating income, measures have been taken to be implemented in almost every field of tourism and the image and security perceptions of tourism destinations have been tried to be changed.

In order to increase tourist security motivation after COVID-19, measures have been taken in all areas of the tourism sector. The measures taken are companies providing transportation services carry passengers within the framework of social distance rules and disinfect the transport vehicles before and after each trip, accommodation establishments attaching importance to disinfect rooms and restaurants by accepting as many guests as the number of people whose hotel occupancy rates will not exceed the social distance norms, tourist guides taking adequate care to comply with personal hygiene rules and social distance rules in order to both ensure their own safety and the safety of tourists, And agencies making tour plans for new and alternative tourism types especially for smaller groups instead of crowded tours. However, it is aimed to increase the security motivation of tourists with pre- and post-travel health checks for the safety of tourists (Özçoban, 2020).

The COVID-19 outbreak has had a serious negative impact on both the tourism industry and tourists. The restrictions imposed have led to the inability to generate income in the tourism sector and also resulted in the complete cessation of activities. On the other hand, it has caused people to reduce or postpone their holiday plans by greatly affecting their travel decisions. In line with the epidemic process and the measures taken after it, it is aimed to turn tourism into an active structure again and to increase tourist security motivations in line with the measures taken in order to achieve this.

2.1 INTRODUCTION

There have been various events that have deeply affected human history and caused the death of millions of people over the generations. Epidemics are among the most common events that cause this condition. In the last quarter of 2019, respiratory tract disease with symptoms, such as fever, cough, and shortness of breath will be among the epidemic diseases that deeply affect the history of humanity. The epidemic disease, which was defined as COVID-19 (new coronavirus) on January 13, 2020, had a very rapid spread effect, causing it to be declared a "pandemic" worldwide after 11 months (WHO, 2020).

This new epidemic has changed the economic, demographic, political, and social balances around the world (Türker and Ertürk, 2020). COVID-19, which seriously damaged the world economy, undoubtedly affected the labor-intensive tourism sector. The tourism sector is in a structure that exists with human labor (Voase, 1995; Yenişehirlioğlu and Salha, 2020). The tourism sector contributes to the employment of millions of people as well as provides foreign currency inflows to countries, affecting their economic development. The cessation of the tourism sector and the closure of tourist businesses have caused millions of people to lose their jobs.

Due to the pandemic, travel restrictions and curfews in line with the measures taken by countries have caused people not to step out of their homes for a very long time. With the control of the pandemic, a period of normalization has been seen, but in the face of this epidemic that can cause deaths, people's travel behaviors and security motivations have developed negatively (Yang and Wong, 2020). Various tasks fall on the tourism industry in the development and restoration of tourist security motivation after the pandemic. The main purpose of this study is to understand the activities carried out by the tourism industry in order to ensure the regeneration of touristic demand by improving the security motivation of tourists during the pandemic process, and to understand the strategies they carry out to regain their economic development potential. This study will be useful for those who need information about the literature and the subject in terms of the development of the security motivation of tourists after the pandemic and the information it will provide in terms of the measures taken by the tourism sector in this process.

2.2 SAFETY MOTIVATION

In the hierarchy of needs determined by Maslow, the concept of security is an important requirement that comes after the physiological needs that are determined as mandatory (Seçilmiş, 2009). Changing world conditions have led to significant developments in people's perception of security. A security problem that may occur on a global scale causes the balance change in many sectors. Today, events that affect the whole world, such as war, terrorism, political conflicts, epidemics, and natural disasters on a global scale cause many sectors to be negatively affected. One of the sectors that are most affected by a possible crisis is the tourism sector. Due to its dynamic structure, the tourism sector is in a structure that can be affected very quickly in the face of a possible crisis (Biggs et al., 2012; Hall, 2010). An international crisis or a negative event that may occur throughout the country can affect people's travel behavior and cause a decrease in tourism demand. Every event that affects the tourism sector also directly affects the security motivations of tourists, as people can cancel their plans if they think they may have anxiety or security problems about a destination they plan to travel to.

Safety motivation is one of the most important concepts affecting the tourism demand of tourists. Planning a safe holiday for people is an indispensable element for the tourism industry. Tourists want their destination of travel to be areas where there are no security problems. Although the touristic attractions, cultural riches, natural beauties, and history of a region affect the travel behavior of people, the priority for tourists is a region that is safe (Kuveloğlu, 2004; Seçilmiş, 2009). The perception of a region as safe by tourists is very effective in the formation of touristic demand. When people plan a vacation, they first head to areas where they can safely travel. In this context, the concept of security is one of the most important concepts that affect people's travel motivation. Various factors affect the security perceptions of tourists about a region. Media news, previous experiences, and perception of security about the place where tourists plan to visit play a decisive role (Seçilmiş, 2009).

One of the most important reasons why the security motivations of tourists are negatively affected by the COVID-19 epidemic is the transmission of the virus from person to person by physical contact. Travels, which are effective in the rapid spread of the virus, have been extremely effective in people's perception of security. The transmission of the virus by humans and travel by the virus carriers negatively affected the safety motivation of tourists (Wilson and Chen, 2020). COVID-19, which affects the whole world today, is an epidemic disease that causes deaths of millions of people. The rapid spread of the virus has caused people to stay away from social environments, and made them stay at home, which is the only place they find safe, and to delay their travels (Akbaba, 2020). The impact of the COVID-19 epidemic all over the world has caused tourists to be affected motivationally and emotionally (Sun et al., 2020). After the outbreak of the epidemic, the security risk perceived by tourists was quite large, leading to the perception of tourism activities as risky and alarming (Reisinger and Mavondo, 2005).

The virus, which creates fear and anxiety in people, has negatively affected the security motivation of tourists and brought tourism activities to a standstill. With the COVID-19 epidemic, tourists have begun to worry about entering social environments that may pose a threat to their safety (Matiza and Slabbert, 2021). The fact that tourists do not find touristic destinations and tourism activities safe in the face of the epidemic that threatens their personal health has caused tourists to worry about traveling. In addition, the fact that tourists do not find transportation services, accommodation businesses, and food and beverage businesses safe in terms of hygiene and health has led to the decrease in touristic demand and the economic loss of businesses in the tourism sector (Kabadayı and Kardeş, 2020).

This new period, in which the security motivation of tourists is significantly affected, has caused many businesses to experience a loss of revenue and also damaged the image perception of the tourism sector (Atay, 2020). The COVID-19 epidemic, in which the security perceptions of tourists toward tourism were significantly affected, led to a significant economic crisis in the tourism sector. Moreover, the fact that tourists cancel their flight tickets because they find it risky has damaged the transportation sector economically. The tourists postponed or canceled the reservations they had made before with the thought that accommodation establishments could pose a threat to health. On the other hand, because food and beverage businesses are also risky, these businesses have started to experience economic problems (Çetinkaya et al., 2020). Tour programs determined by travel agencies have been canceled due to tourists avoiding crowded environments. Cancellation of tours led tourist guides to become unemployed.

The tourists' concerns about hygiene, finding social environments risky, and not wanting to be in crowded areas have affected their travel behaviors and caused a decrease in security motivation. The new era brought by the pandemic required various measures to be taken in order to increase the security motivation of tourists. It is expected that the security motivation of tourists will increase and the demand for tourism will increase if the businesses in the tourism sector comply with the hygiene and safety rules in this new period (Türker and Ertürk, 2020).

2.3 TOURISM SECTOR AFTER THE COVID-19 PANDEMIC

In late December 2019, as a result of research conducted in the seafood and animal market in Wuhan, China, COVID-19 (new coronavirus) pandemic emerged on 13 January 2020 (TR. Ministry of Health COVID-19-Guide, 2020). The coronavirus, which spreads by the scattering of droplets originating from the respiratory tract, such as sneezing and coughing, is transmitted by people in contact with each other. The coronavirus first appeared in the city of Wuhan, but soon spread within China. The virus, which spreads very quickly, spread to many countries in a short time due to the high population of China and international travels (Velavan and Meyer, 2020).

The coronavirus first spread to China, USA, Italy, Spain, Germany, France, Iran, United Kingdom, Turkey, Switzerland, Belgium, Netherlands, Canada, Austria, Korea and later spread to 210 countries in total (WHO). The virus spreads very quickly as a result of people's contact with each other has caused the death of many people around the world today. Not knowing when the pandemic will end and showing a rapid spread effect, several countries have started to take various measures to prevent the spread of the coronavirus.

After the outbreak of the epidemic, countries primarily closed their borders and restricted the entry and exit to the country. With the prohibition of international travel, the entry of foreigners into the country was completely stopped. The closure of the workplaces led to the cessation of production and thus to an economic crisis. In addition, environments that may pose risks, especially social environments, such as schools, mosques, churches, museums, parks, restaurants, hotels, and cafes where the virus can spread rapidly, have been closed (Kuipers et al., 2021). Collective activities, such as concerts, excursions, sports competitions, and entertainment activities where people come together were canceled and curfews were implemented in order to prevent people from staying away from home and contacting others in this process (Kıvılcım, 2020). One of the measures taken for people who are caught in the COVID-19 pandemic is the recommendation to stay at home for 14 days and enter quarantine.

In this new developing process, the spread of the COVID-19 pandemic all over the world has seriously affected the economies of the countries. COVID-19 has caused countries to experience economic losses and people have become unemployed. The negative effects of many sectors that provide foreign exchange inflows to countries and provide advantages in economic development have led to income losses (Atay, 2020).

Tourism, which has a significant impact on the development of countries, is a dynamic sector that is affected very quickly by global events. Events such as war, terrorism, and natural disasters all over the world affect the tourism sector (Acar, 2020). Another situation that the tourism sector is heavily affected is by epidemics. One of the sectors most affected by COVID-19, which has affected the whole world, has been the tourism sector. The tourism sector, which provides foreign currency inflows to countries and has a significant impact on economic development, has been seriously affected by the restriction of travel during this process, and people do not find it safe to travel and started staying at home and were not participating in tourism activities (Eryüzlü, 2020).

According to a report of the United Nations World Tourism Organization examining the effects of the COVID-19pandemic on the tourism sector, the rate of tourists decreased by 22% in the first months of 2020 at the international level. It has been stated that by the end of 2020, there may be a 60–80% decrease in the tourist rate (UNWTO, 2020). According to the UNWTO report, a reduction in tourist travel in these situstions could lead to the following problems:

- Loss of 850 million to 1.1 million international tourists.
- Loss of export income between 910 billion and 1.2 trillion dollars from tourism sector.
- Loss of employment between 100 and 120 million in the tourism sector.

The fact that tourists do not participate in tourism activities due to travel barriers and security concerns will reduce the rate of tourists. This situation will significantly affect not only the tourism sector but also the economy of countries that provide significant income from the tourism business (Alaeddinoğlu and Rol, 2020).

When the crisis caused by COVID-19 is evaluated in terms of tourists, the fact that touristic destinations are not safe due to social and crowded environments, and the security concerns of the tourists due to the epidemic have caused tourists to cancel their reservations. This situation has resulted in the closure of transportation businesses, accommodation businesses, and food and beverage businesses, such as restaurants and cafes, bringing domestic and international tourism activities to a standstill, leading to a decrease in tourist expenditures and thus led to a decrease in tourism demand (Bakar and Rosbi, 2020).

Within the scope of COVID-19, the pandemic process negatively affected the tourism sector and the countries that provide income from the tourism sector, also negatively affecting the transportation sector, tour operators, tourism agencies, tourist guides, accommodation, and food and beverage operators and people working in these fields. Within the scope of the normalization process, in order to mitigate the damage that may arise from COVID-19 in the tourism sector, it is aimed to minimize the negative effects of the epidemic on the tourism sector by taking precautions in various areas, especially to ensure social distance rules and hygiene rules (Aydın and Doğan, 2020).

2.4 THE ROLE OF THE TOURISM INDUSTRY IN TOURIST SAFETY MOTIVATION AFTER THE PANDEMIC

COVID-19, starting in China, has taken the whole world under its influence in a short time. The death of millions of people due to the epidemic and the inability of people to leave their homes for a long time have led to various anxiety and anxiety problems. Safety motivation, which significantly affects people's travel behavior, has become a more important criterion with the epidemic (Ranasinghe et al., 2020). There has been a significant decrease in the security motivation of the tourism sector due to the effect of spreading faster in crowded and social environments. Countries have started to take various measures to reduce the spread of the epidemic. In the beginning, the closing of the borders of the countries, travel restrictions, intercity transportation bans, closure of social areas, quarantine practices, and curfews are the main measures taken. The thought, anxiety and worry of the health risk posed by the epidemic on tourists also affected tourism demand (Wachyuni and Kusumaningrum, 2020). The tourism sector has started to take various measures within the enterprise in order to ensure the regeneration of touristic demand and to increase the security motivation of tourists. The normalization process that the whole world is in includes developments to improve the security perceptions and motivations of tourists, which are effective in participating in tourism.

With the "new normal" period, the decisions taken within the accommodation enterprises in order to increase the security motivation of the tourists and to increase the economic income of the tourism sector are listed as follows (Çetinkaya et al., 2020):

- Accommodation establishments provide written information about the measures and practices taken before the guest is accepted, and besides this, they place signs to pay attention to social distance in crowded areas.
- Accommodation businesses accept guests at the capacity determined within the social distance rules and place thermal cameras and noncontact thermometers at the entrances of the facilities.
- During the stay, the necessary equipment is provided for the guests to comply with the personal hygiene rules, and the information about the chronic diseases and whether the guests have had COVID-19 in the last 14 days are recorded.
- In the open buffets in the accommodation establishments, the service is carried out by following the social distance rules of the kitchen staff. However, the distance between the dining tables is 1.5 m and the distance between the chairs is 60 cm.
- Health checks of the operating personnel are carried out regularly, and it is ensured that the personnel are trained in hygiene and they pay attention to the guests and their personal safety.
- Accommodation businesses take care to disinfect all areas after customers' stays are over.

The tourism sector aimed to minimize all negative factors that could threaten the health of the guests by keeping the hygiene rules at the highest level in order to provide the security motivation of the tourists during COVID-19. In this context, hygiene rules in accommodation businesses, control of business entrances and exits, social distance practices, sterilization of general areas (bedroom, distance between sun loungers at the beach and pools, swimming pool, gym, animation hall, eating–drinking area, and kitchen), personnel-oriented measures, contactless payment, hygiene measures and other measures taken for the safety of guests have started to be implemented (Yenişehirlioğlu and Salha, 2020).

With the entry of the new normalization period, tourist destinations have reopened and various rules have been determined (Farzanegan et al., 2021). In line with the determined rules, it is necessary for tourists and officials to use masks, to comply with the social distance rules, and to pay attention to the hygiene rules. Another area where various measures are taken that can affect the security motivations of tourists and reduce their concerns is the food and beverage sector. Food and beverage businesses should design and arrange tables and chairs in accordance with the social distance guidelines to spread customers and prevent them from infection. Disinfectant stands should be placed, especially at the entrances and exits of the enterprise. Eating areas should be disinfected at regular intervals. Business personnel should use hygienic equipment (apron, mask, visor, and gloves) (Çetin and Coşkuner, 2021). The drinks to be served to the guests should be closed, and the service equipment should be brought to the tables in a packaged way and should be disposable (Dündar et al., 2020). However, care should be taken to ventilate the environment. Within the scope of the measures taken in the food and beverage establishments, it is aimed to positively affect the security motivation of the tourists and to minimize their anxieties and worries.

In the transportation sector, which has an important place in the tourism sector, it has been determined to pay attention to the social distance rule and to travel by leaving empty seats among the passengers. In addition, the use of masks is mandatory throughout the journey and the hygiene equipment that the passengers may need is provided. After the trip, the vehicle is sanitized and made sterile for the next trip. One of the most important tasks in improving the perception of tourist security falls on tourist guides. Since the tourist guides will be with the tourists throughout the tour, they will have a serious impact on the security perception of the tourists by giving the necessary warnings to protect the social distance, hygiene measures, and informing the tourists (Türker and Karaca, 2020).

The increase in the level of anxiety and stress in tourists with the COVID-19 pandemic required measures to be taken in all areas of the tourism sector. The security motivation, which was significantly effective in reducing the motivation to travel in people with the control of the epidemic, provided incentives for tourists to participate in tourism as a result of the measures taken. The measures taken to improve the security motivations for tourism during the COVID-19 pandemic have also been effective in the development of the tourism sector. The measures taken to eliminate the uneasiness of tourists at the point of participating in tourism activities after the epidemic influence the tourists to travel more safely and reduce the rate of catching the coronavirus (Aydın and Doğan, 2020). Since the measures implemented to increase the security motivation will create an image that the tourists have not experienced before, this will affect the tourists to gain new experiences and increase their motivation to travel. Measures that increase the security motivation of tourists after the pandemic will increase their participation in touristic activities and will provide an opportunity to increase the purchase of touristic goods and services (Özdemir, 2020). As people spend almost all of their time at home during the epidemic, the measures taken in the tourism sector will also affect tourists' visit intentions and recommendation intentions. This situation may affect the recovery of the economic development power provided by the tourism sector over time.

2.5 CONCLUSIONS

The COVID-19 pandemic, which emerged in December 2019, has taken the whole world under its influence in a short time. The rapid spread of the virus caused a worldwide pandemic to be declared. Countries have taken various measures against the virus epidemic that spreads as a result of people's contact with each other. First, quarantine practices and curfews were introduced after the borders were closed and international entrances and exits were restricted. Closing social areas, paying attention to social distance rules in areas where people can find together, and making the use of masks compulsory are among the other measures taken.

The restrictions and bans imposed due to the rapid spread of the epidemic have negatively affected the countries in terms of social and economic aspects. Tourism, which makes an economic contribution to the development of countries and the increase in employment, has been one of the sectors most affected by this process (Kalyankar and Patil, 2020). The tourism sector, which was greatly affected by the negative events that took place on a global scale, started to lose income in this process. The economic losses in the tourism sector have caused the employment rate of the tourism sector to decrease and people to be unemployed.

The COVID-19 pandemic has caused a lot of damage, both economically and sociopsychologically, all over the world. Countries that provide development with the income obtained from tourism have entered a period in which significant economic problems will be experienced. On the other hand, the fact that the epidemic caused deaths caused anxiety and stress in people's travel perceptions. Over time, with the epidemic being brought under control, it started a new normalization process. This new normalization process has brought along various measures to revive the tourism sector. In this context, various measures have been taken for accommodation businesses, food and beverage businesses, transportation businesses, travel agencies, and tourist guides in order to regain the security motivation of tourists. With the measures taken, it is necessary to minimize the situations that may threaten the health of the tourists, to comply with the social distance rules, and to pay attention to the hygiene rules. People who have been at home since the beginning of the epidemic have negative thoughts about touristic areas and social environments. The tourism sector, which wants to minimize the worries and concerns of the tourists who have been affected by the epidemic, aims to increase the touristic demand again as a result of the measures taken. It is assumed that this situation will positively change the perception of security created by the COVID-19 pandemic in tourists, as well as will increase the purchase rate of touristic goods and services, thereby reviving the demand for tourism.

KEYWORDS

- **safety**
- **motivation**
- **tourism**
- **tourist motivation**
- **pandemic disaster**

REFERENCES

Acar, Y. Yeni Koronavirüs (COVID-19) Salgını ve Turizm Faaliyetlerine Etkisi. *Güncel Turizm Araştırmaları Dergisi* **2020,** *4* (1), 7–21.

Akbaba, M. COVID-19 Salgını Kapsamında Turist Yorgunluğunun Turistlerin Satın Alma, Tavsiye Etme ve Daha Fazla Ödeme Niyetlerine Etkisinin Belirlenmesi. *Int. J. Soc. Sci. Educ. Res.* **2020,** *6* (2), 225–240.

Alaeddinoğlu, F.; Rol, S. COVID-19 Pandemisi ve Turizm Üzerindeki Etkileri. *Yüzüncü Yıl Üniversitesi Sosyal Bilimler Enstitüsü Dergisi* (Salgın Hastalıklar Özel Sayısı) **2020,** 233–258.

Atay, L. Kovıd-19 Salgını ve Turizme Etkileri. *Seyahat ve Otel İşletmeciliği Dergisi* **2020,** *17* (1), 168–172.

Aydın, B.; Doğan, M. Yeni Koronavirüs (COVID-19) Pandemisinin Turistik Tüketici Davranışları ve Türkiye Turizmi Üzerindeki Etkilerinin Değerlendirilmesi. *Pazarlama Teorisi ve Uygulamaları Dergisi* **2020,** *6* (1), 93–115.

Bakar, N. A.; Rosbi, S. Effect of Coronavirus Disease (COVID-19) to Tourism İndustry. *Int. J. Adv. Eng. Res. Sci.* **2020,** *7* (4), 189–193.

Biggs, D.; Hall, C. M.; Stoeckl, N. The Resilience of Formal and Informal Tourism Enterprises to Disasters: Reef Tourism in Phuket, Thailand. *J. Sustain. Tour.* **2012,** *20* (5), 645–665.

COVID-19: Operational guidance for maintaining essential health services during an outbreak, interim guidance https://www.who.int/publications-detail/COVID-19-operationalguidance-for-maintaining-essential-health-services-during-an-outbreak (Accessed 02 Oct 2022).

Çetin, A.; Coşkuner, M. New Practice in Accommodation Facilities After COVID-19: Safe Tourism Certification Program. *J. Hosp. Tour. Iss.* **2021,** *3* (1), 16–22.

Çetinkaya, Ö.; Özer, Y.; Altunel, G. K. COVID-19 Sürecinde Turistik Seyahat Risk Algısı ve Hijyen-Güvenlik Algısının Değerlendirilmesi. *Tour. Recreat.* **2020,** *2* (2), 78–83.

Dündar, Y.; Silik, C. E.; Ilgaz, B. Antalya'da Yer Alan Konaklama Tesisleri Yöneticilerinin COVID-19 Kapsamında Alınabilecek Tedbirler Üzerine Görüşleri. *İşletme Araştırmaları Dergisi* **2020,** *12* (4), 3776–3794.

Emekli, G. Coğrafya, Kültür Ve Turizm: Kültürel Turizm. *Ege Coğrafya Dergisi* **2006,** *15* (1–2), 51–59.

Eryüzlü, H. COVID-19 Ekonomik Etkileri ve Tedbirler: Türkiye'de "Helikopter Para" Uygulaması. *Ekonomi Maliye İşletme Dergisi* **2020,** *3* (1), 10–19.

Farzanegan, M. R.; Gholipour, H. F.; Feizi, M.; Nunkoo, R.; Andargoli, A. E. International Tourism and Outbreak of Coronavirus (COVID-19): A Cross-Country Analysis. *J. Travel Res.* **2021,** *60* (3), 687–692.

Hall, C. M. Crisis Events in Tourism: Subjects of Crisis in Tourism. *Curr. İss. Tour.* **2010,** *13* (5), 401–417.

Kabadayı, M.; Kardeş, N. Kovid-19'un (Koronavirüsün) Yerli Turist Davranışı ve Seyahat Eğilimlerine Etkileri. *Türk Turizm Araştırmaları Dergisi* **2020,** *4* (4), 3703–3719.

Kalyankar, M. D.; Patil, P. Impact of COVID-19 Pandemic on the Tourism Sector. *Purakala (UGC Care J.)* **2020,** *31* (8), 611–617.

Kaygısız, A. D. COVID-19 Sonrası Türk Turizm Sektöründe Oluşabilecek Fırsatlar ve Riskler. *Bartın Üniversitesi İktisadi ve İdari Bilimler Fakültesi Dergisi* **2021,** *12* (23), 79–95.

Kıvılcım, B. COVID-19 (Yeni Koronavirüs) Salgınının Turizm Sektörüne Muhtemel Etkileri. *Uluslararası Batı Karadeniz Sosyal ve Beşeri Bilimler Dergisi* **2020,** *4* (1), 17–27.

Kuipers, N.; Mujani, S.; Pepinsky, T.Encouraging Indonesians to Pray from Home During the COVID-19 Pandemic. *J. Exp. Polit. Sci.* **2021,** *8* (3), 211–222.

Kuveloğlu, D. Turizm@gelecek.tr; Elips Kitap: Ankara, 2004.

Matiza, T.; Slabbert, E. Tourism is Too Dangerous! Perceived Risk and the Subjective Safety of Tourism Activity in the Era of COVID-19. *Geo J. Tour. Geosites* **2021,** *36*, 580–588.

Özçoban, E. Koronavirüs' ün (COVID-19) Turizm Sektörü Üzerindeki Etkileri ve Türkiye'nin Kırsal Turizm Potansiyeli Üzerine Bir Analiz. *Electron. Turk. Stud.* **2020,** *15* (4).

Özdemir, M. A. COVID-19 Salgını Sonrası Alınan Önlemlerle Turizm: Muhtemel Senaryolar. *J. Recreat. Tour. Res./JRTR* **2020,** *7* (2), 222–238.

Ranasinghe, R.; Damunupola, A.; Wijesundara, S.; Karunarathna, C.; Nawarathna, D.; Gamage, S.; Idroos, A. A. Tourism After Corona: Impacts of COVID 19 Pandemic and Way Forward for Tourism, Hotel and Mice Industry in Sri Lanka. *Hotel and Mice Industry in Sri Lanka, April 22, 2020.*

Reisinger, Y.; Mavondo, F. Travel Anxiety and Intentions to Travel Internationally: Implications of Travel Risk Perception. *J. Travel Res.* **2005,** *43* (3), 212–225.

Seçilmiş, C. Turistlerin Kişisel Değişkenlerinin Güvenlik Algılamalarındaki Rolü. *Elektronik Sosyal Bilimler Dergisi* **2009,** *8* (30), 152–166.

Sun, J.; Zhang, J.H.; Zhang, H.; Wang, C.; Duan, X.; Chen, M. Development and validation of a tourism fatigue scale. *Tour. Manage.* **2020,** *81*, 104121.

Sağlık Bakanlığı, T. C. Halk Sağlığı genel Müdürlüğü, COVID-19 (2019-n CoV Hastalığı) Rehberi (Bilim Kurulu Çalışması). T.C. Sağlık Bakanlığı, 25 Şubat 2020. https://covid19bilgi.saglik.gov.tr/depo/rehberler/COVID-19_Rehberi.pdf (Accessed 07 Oct 2022).

Türker, A.; Karaca, K. Ç. Yeni Tip Koronavirüs (COVID-19) Salgını Sonrası Turizm ve Turist Rehberliği. *Turist Rehberliği Nitel Araştırmalari Dergisi* **2020,** *1* (1), 1–19.

Türker, G. Ö.; Ertürk, N. COVID-19'un Konaklama İşletmelerine Etkileri: Yöneticiler Perspektifinden Bir Değerlendirme. *Turizm Ekonomi Ve İşletme Araştırmaları Dergisi* **2020,** *2* (2), 89–101.

UNWTO. International Tourist Numbers Could Fall 60–80% in 2020, Unwto Reports. https://www.unwto.org/news/COVID-19-international-touristnumbers-could-fall-60–80-in-2020 (Accessed 02 Jan 2022).

Velavan, T. P.; Meyer, C. G. The COVID-19 Epidemic. *Trop. Med. İnt. Health* **2020,** *25* (3), 278.

Voase, R. *Tourism: The Human Perspective*; Hodder and Stoughton, 1995.

Wachyuni, S. S.; Kusumaningrum, D. A. The Effect of COVID-19 Pandemic: How Are the Future Tourist Behavior? *J. Educ. Soc. Behav. Sci.* **2020,** 67–76.

Wilson, M. E.; Chen, L. H. Travellers Give Wings to Novel Coronavirus (2019-nCoV). *J. Travel Med.* **2020,** 1–3.

World Health Organization. Rolling Updates on Coronavirus Disease (COVID-19). https://www.who.int/emergencies/diseases/novel-coronavirus2019/events-as-they-happen (Accessed 15 May 2020).

Yang, F. X.; Wong, I. A. The Social Crisis Aftermath: Tourist Well-Being During the COVID-19 Outbreak. *J. Sustain. Tour.* **2020,** *29* (6), 859–878.

Yenişehirlioğlu, E.; Salha, H. COVID 19 Pandemisinin Türkiye İç Turizmine Yansımaları: Değişen Talep Üzerine Bir Araştırma. *İstanbul Ticaret Üniversitesi Sosyal Bilimler Dergisi* **2020,** *19* (37), 355–368.

CHAPTER 3

The Recovery of the Tourism Industry from Pandemics and Disasters

HATICE SARI GÖK[1] and FATMANUR KÜBRA AYLAN[2]

[1]*Department of Travel, Tourism and Entertainment Services, Yalvaç Vocational School, Isparta University of Applied Science, Isparta, Turkey*

[2]*Recreation Management Department, Tourism Faculty, Selçuk University, Konya, Turkey*

ABSTRACT

As the tourism industry is developed upon quite sensitive balances, it is influenced by several incidents in a negative and positive manner to a national and international extent. Particularly, the pandemic and natural disasters such as floods, hurricanes, forest fires, droughts, and earthquakes that affected them hindered the development and improvement of the tourism industry. In recent days, the pandemic, which has affected a lot of countries around the world, has influenced the tourism industry as well as many other sectors in national and international dimensions. Similarly, as natural disasters also have the potential to harm tourism establishments, touristic attractions, and tourists, the situation of the destination and the country experienced and the crisis environment influence tourism for certain. From this point of view, the pandemics and disasters experienced in the study were discussed, and the effects of pandemics and disasters on the tourism industry were examined. Suggestions for the recovery of the tourism industry from pandemics and disasters were presented.

Dynamics of the Tourism Industry: Post-Pandemic and Post-Disaster Perspectives and Strategies.
Gül Erkol Bayram and Anukrati Sharma (Eds.)

3.1 INTRODUCTION

Various pandemics occurred throughout the world history, and these epidemics/pandemics gave rise to the deaths of several people. The fact that pandemics bring about health problems along with economic problems leads to various sectors being affected. Without any doubt, one of the sectors being affected by epidemics is the tourism sector. Especially the tourism sector is influenced negatively in touristic regions, where epidemics are spread. For instance, the hand, foot, and mouth epidemic that emerged in the United Kingdom in February, 2001 is one of the most communicable diseases for animals and human beings. Since this disease can be infected through direct contact with infected animals or their body fluids (urine, milk, saliva etc.), and through air, it can be said that the tourism section is influenced as a result of infection. Another epidemic having an influence on the tourism industry is SARS (severe acute respiratory syndrome). SARS and the bird flu epidemic damaged the image that Asia is a secure tourist destination. China, Hong Kong, Singapore, and Vietnam were affected by the SARS epidemic. The tourism industry was severely affected in these countries because of the epidemic, and the industry came to the point of downfall. Finally, the COVID-19 pandemic, which occurred in Wuhan, city of China, in 2019 took hold of the whole world (Khalilzadeh, 2020).

Natural disasters are disasters that happen unexpectedly and accompany many destructive effects, such as floods, fires, earthquakes, tsunamis, and volcanic eruptions. There are some sets of problems occurred in meeting the needs of the local community as many resources in the region where natural disasters occurred have disappeared. Together with this, it is known that many sectors in the region were affected by the disaster. Notably, if the natural disaster happened in a tourism region, the tourism industry in the region would also be affected beyond doubt. Touristic demand for the region may also decrease as a result of the damage to the security perception of tourists or the image of the region. For example, tourism reservations were cancelled to a great extent following the Gölcük–Adapazarı earthquake on August 17, 1999, in Turkey, and tourism industry of Turkey was affected by this disaster in no small measure. Within the light of sources, the disasters and pandemics experienced throughout the world are analyzed in this study, and the effects on the tourism industry are tried to be set forth. At the same time, the studies and implementations conducted to revive the tourism industry after epidemics and disasters occurred are explained through examples. Furthermore, suggestions on the revival of the tourism industry following pandemics and natural disasters are also provided in the study. In-depth document searching method is used in the study, and it is considered that it will contribute to the tourism industry and the literature.

3.2 PANDEMIC, CONTEXT, AND DEFINITIONS

Contagious diseases are known as epidemics. Pandemics are contagious diseases that spread to vast geographies, sometimes to a continent or the whole world and lead to death, and diseases in humans and animals (Aslan, 2020: 35). It is known that various diseases occurred throughout world history. For example, diseases such as Plague of Amwas, Black Plague, Spanish Flu, Cholera, Hong Kong flu, and Variola have happened in the history and given rise to deaths of many people. When the influences of these epidemics on the world are analyzed, the Black Plague, known as the great epidemic plague, started in China and the southern west of Asia, reached Europe between 1347 and 1351, and caused the deaths of 200 million people in the 14th century. The Spanish Flu, a fatal sub-type of the H1N1 virus, occurred between 1918 and 1920 and resulted in the deaths of approximately 100 million people in 18 months. Another example of an epidemic is the Hong Kong Flu, which emerged in Hong Kong for the first time, although it is known as Asia Flu or China Flu. It is a flu epidemic having given rise to the deaths of about one million people between 1968 and 1969 (Aslan, 2020). Information on more recent outbreaks SARS, swine flu, avian flu, and MERS and ongoing pandemics HIV/AIDS and COVID-19 is listed below (Table 3.1) (Karcıoğlu, 2020);

World Health Organization (WHO) published a report in 2018 named "Management of Epidemic Diseases". In this report, it was brought to drawn attention that the possibility of a new Ebola or flu epidemic is not low, but the timing of it cannot be estimated. Following a short period of time after the report of WHO, as a result of the research in a group of patients with symptoms of respiratory diseases such as fever, cough, and difficulty breathing at the end of 2019, the information on the identification of a new type of coronavirus on January 13, 2020, proved the foresight of WHO. Named as COVID-19 by WHO following the identification of the disease on February 11, 2020, this new type of corona virus was announced as a pandemic one month later on March 11, 2020, due to its prompt spread (Alaeddinoğlu and Rol, 2020: 234). States took a variety of precautions and began to apply implement them after the announcement of the pandemic to curb the disease.

TABLE 3.1 Comparison of the Specifications of Certain Epidemics and Pandemics Took Place in the World in Different Periods of Time.

Disease	Region	Year	Mortality rate (fatality)
Influenza Epidemic "Spanish Flu"	Spain and Whole World	1918–1920	0;2–0;5%
HIV-AIDS	Started in Africa-Macaque and whole world	1984-continuing	3;9 % (person/year)
"Severe acute respiratory syndrome" (SARS)	Wildcats and people in Far East	2002–2003	10%
H1N1 (Swine flu)	Whole World	2009–2010	0;1–0;2%
Seasonal influenza	33 countries (countries with high mortality rate, 57% of world population)	Wide range (1999–2015)	01 and 64/ 100 000 (under 65) 3 and 223/ 100 000 (over 65)
"Middle East Respiratory Syndrome" (MERS)	Saudi Arabia	2012- (Last diagnosis was made in 2019.)	35–40%
COVID-19	Started in China and whole world	2019-Continuing.	(Estimated) 1–3%

Source: Adapted from Karcıoğlu (2020).

The COVID-19 influenced the whole world with its economic, political, demographic, and social effects. The pandemic inflicted a heavy blow on the tourism industry as well as all sectors of the world economy. Limitations such as countries' sealing the borders, ceasing airline transport, and taking quarantine precautions bring the travel and tourism sectors to a halt. The closure of touristic enterprises not only caused a loss in the consumption economy but also induced the unemployment of millions of employees working in tourism enterprises.

3.2.1 EPIDEMICS IN TOURISM INDUSTRY

The crisis arising from epidemic diseases differs in terms of its influences on the tourism industry compared to the durations of terrorist incident, political crises, and natural disasters. According to the searches, the duration for destinations to get rid of the crisis of epidemic diseases varies between 12 and 14 months on average. Moreover, the progress of the recovery period in the tourism industry has a parallel course with the number of cases of epidemic diseases. For instance, international arrivals to Sierra Leone even 35 months after the Ebola epidemic remained under 50% of the number of visitors before the epidemic. Hand, foot, and mouth disease that appeared in England in 2001 hampered the growth of the tourism industry, and growth rates remained at pre-epidemic level for two years. Similarly, a sharp decrease took place in demand for tourism in Hong Kong as a result of SARS in the two-quarters period (Khalilzadeh, 2020: 2).

In our day, epidemic diseases spread to other regions and countries from where they started through air/sea transportation, particularly through developed transportation routes. Hand, foot, and mouth disease, severe acute respiratory syndrome (SARS), bird flu, swine flu, Middle East respiratory syndrome (MERS), Ebola, and COVID-19 can be given as examples of the epidemics affecting the tourism industry. The most significant negative effect of epidemics on the tourism industry is the decrease in demand directed at the destination. Information on the epidemics affecting the tourism industry is given in the following:

Hand, Foot, and Mouth Disease: It emerged in February 2001 in the United Kingdom. Hand, foot, and mouth disease is one of the most contagious diseases, both for animals and people. The animals infected by this disease spread it rapidly. It spreads as a result of contact with infected animals or bodily fluids of them like milk, saliva, or urine, as well as airborne transmission

(Grubman and Baxt, 2004: 466). Hand, foot, and mouth disease led to a 9.61% decrease in the number of tourists coming to the United Kingdom in 2001 compared to previous year. The number of tourists before the epidemic could only be reached in 2004 (Çeti and Ünlüönen, 2019: 116).

SARS: It is a severe acute respiratory syndrome that appeared in live-stock market workers for the first time in November 2002 in China. Various cities in China were announced as epidemic areas after a short while after the emergence of the first patient. Epidemic SARS was later seen in Southern East Asia and other regions of the world and spread to 26 countries on five continents (Zeng et al., 2005, 307–308; Wilder-Smith, 2006: 53). 774 people died of the disease, and 8098 people were infected in total between November 1, 2002 and July 31, 2003. The countries with highest cases are ranked as China (5327), Hong Kong (1755), Taiwan (346), Canada (251), and Singapore (238) (Çeti and Ünlüönen, 2019: 177).

Despite the fact that SARS emerged as a local epidemic, it was turned into a global pandemic by a traveller. A businessman who travelled from China to Hong Kong transmitted the disease to Vietnam on February 23, 2003. As a result of the emergence of this "mysterious disease" in a hospital in Vietnam, World Health Organization (WHO) issued a global alarm on March 12, 2003 (Wilder-Smith, 2006: 54). Furthermore, as a result of the fact that international tourists visited the area infected by the virus lead to transmission of the disease to their countries, the World Health Organization (WTO) declared Hong Kong and Guangzhou cities as high-risk cities (Mason et al., 2005). Known as "the pearl of Asia", Hong Kong was labelled "City with Mask" as the people walking in the streets were wearing masks after SARS disease (Au et al., 2005). Many countries limited the entrance of Chinese to their countries as well as transportation with China and Hong Kong.

Even though the effects of epidemic SARS did not spread equally in a sectoral or geographical sense, it resulted in a significant regression in the tourism industry in the first half of 2003. There was an increase in uncertainty in the demand for global journeys (Mason et al., 2005). Moreover, tourists' abandoning distant travels and turning towards close destinations caused a growth in the negative effects of SARS on hotels and travel agencies. Though not having statistically clear figures, all kinds of enterprises in the regions where the epidemic occurred were affected by this negation (Gu and Wall, 2006). It is specified that SARS caused a decline in the number of tourists coming to Taiwan substantially but there were various attempts after the epidemic to gain the trust of tourists, and the number of tourists coming to the country exceeded 3 million in 2005 for the first time after the epidemic (Wang, 2009: 78–79).

MERS: The respiratory syndrome, rooted in the Middle East, is a corona virus origin disease. It was reported in 2010 in Saudi Arabia for the first time. It particularly influenced Arabian Peninsula as of September 2012. It was mostly observed in 2494 people in a relatively limited geography in Saudi Arabia between 2010 and 2019 (November) and gave rise to deaths of 858 people (Khalilzadeh, 2020: 3). The first cases outside the Arabian Peninsula were seen among those travelling to Germany, Greece, Italy, France, the Netherlands, and England. Apart from Europe, it was seen in countries such as Egypt, Malaysia, Tunisia, and the Philippines (Pavli et al., 2014). One of the countries affected by the pandemic was South Korea. There was a loss in tourism income in South Korea of about 2.6 billion dollars. Likewise, there was a decrease in the number of tourists at a rate of approximately 37% from June to September 2015. One of the most important reasons for this situation was explained by the fact that tourists cancelled their travels as they feared the disease (Joo et al., 2019).

Ebola: The first incidence of an Ebola epidemic case was in Guinea in December 2013. According to the World Health Organization, there were 28,646 cases in total until March 27, 2016, and 11323 people lost their lives. The countries with the highest number of cases were Sierra Leone (14124), Liberia (10,675), Guinea (3811), Nigeria (20), Mali (8). 11,322 of the deaths took place in these five countries (Çeti and Ünlüönen, 2019: 121). Ebola disease having been effective on the African continent affected Gambia's tourism in a negative manner (Novelli et al., 2018). Similarly, the Ebola epidemic caused serious harm to the tourism industry of the whole continent in 2014, even in destinationswhere Ebola did not occur in Western Africa (Mizrachi and Fuchs, 2016: 59).

Bird Flu: It emerged in the gooses in the south of China in 1996 for the first time. It is a disease known as avian flu that is transmitted from migratory birds to poultry. Having fatal influences on poultry, bird flu then infected human beings. In 1997, it infected people in Hong Kong and then had influence over many countries like China, Vietnam, Thailand, Cambodia, Indonesia, Egypt, Djibouti, and Azerbaijan. 6 of the 18 people who were infected by the virus lost their lives. There were 238 cases in total from the first incidence of the disease until January 19, 2006, and 139 people lost their lives (Kiper et al., 2020: 532).

Swine Flu: The swine flu virus was first seen in people in April, 2009 in the United States of America. Swine flu infects people by sitting close to infected people, talking, walking, or sneezing, coughing, spitting, or coming into contact with infected areas (Haque and Haque, 2018: 92). According to the data of World Health Organization, 13,398 swine flu cases were recorded in 48 countries until May 27, 2009, and 95 people lost their lives. The countries with the highest case rates were the United States of America (6764), Mexico (4541), Canada (921), Japan (360), Spain (138). The mortality rate was declared as Mexico (83), the United States of America (10), Canada (1), and Costa Rika (1) (Çeti and Ünlüönen, 2019: 119–120). It affected the international demand for tourism to a great extent in 2008 in the United Kingdom (Page et al., 2006).

COVID-19: The disease originating from a virus was accepted to be limited to China and deemed as "epidemic" by WHO announced it as a global "pandemic" on March 11, 2020. Unlike other diseases, the most important feature of COVID-19 disease is its rapid and widespread. Social isolation, lockdowns, quarantine measures, obligation to wear masks, and limitations on travel became the first precautions taken to prevent the spread of COVID-19 disease. These implementations affected production and consumption in a negative manner from an economic perspective. It resulted in a slowdown and cease of economic activities and led to a global economic recession (TURSAB, 2020).

Obscurity in infection and treatment directly affected the tourism industry due to obligatory travel measures. Majority of journeys were cancelled or postponed following the emergence of COVID-19 in China (Çetin and Göktepe, 2020: 90). Along with affecting many sectors, tourism industry was also negatively affected by COVID-19. The effects of pandemic COVID-19 on tourism industry are dealt with in detail in the following.

3.2.2 INFLUENCE OF PANDEMIC COVID-19 ON TOURISM INDUSTRY

The tourism industry is expected to be among the first sectors to be affected and the last to recover from the pandemic because it is partially a luxurious means of consumption with its deferrable structure. The tourism industry almost came to a stopping point throughout the world due to wider and more harsh travel limitations compared to previous diseases, and many countries limited travel except for obligatory situations. A lot of travel agencies, tour operators, and airline companies ceased their activities. In addition to this, accommodation management also had to discontinue their activities. A wide range of limitations affecting the entertainment and leisure time sectors gave rise to distortions in the global supply chain (Boone et al., 2020).

Together with the fact that COVID-19 is a major health problem for the world, it led to various social and economic problems. It affected global commerce, and, particularly, the tourism industry deeply. Situations such as the closure of borders at the global level and cancellation of flights, the closure of some hotels, and limitations on travel at the national and local levels harmed the tourism industry in all countries (Aydın and Doğan, 2020: 95). According to April 28, 2020, dated report published by the World Travel and Tourism Council (WTTC), influence of COVID-19 pandemic on the global economy is eight times greater than the effect of the 2008 global economic crisis. The reason for that is regression of employment in the tourism industry in 2020 at the rate of 31% (unemployment of 100 million people) and a decrease in tourism income at the rate of 30% (2.7 trillion American dollar). WTTC defines this situation as "unique losses for 2020" (WTTC, 2020).

According to the data of the World Tourism Organization (UNWTO), approximately 1.5 billion international travels were actualized in 2019. International tourist arrivals decreased at a 70% rate due to the proliferation of travel limitations and a decrease in visitors' demand in 2020 when the pandemic started to spread and its effects spread to the whole world (UNWTO, 2020). The economic problems that pandemic COVID-19 caused resulted in the employment termination of workers in the tourism industry, generally by paying little or no compensation (Baum and Hai, 2020). Massive employment loss continuing because of pandemic COVID-19 enlarges the injustice and problems of low income, seasonal worker applications, and deprivation of social rights that tourism industry workers already face (Benjamin et al., 2020). The ones who are most affected by these problems are female workers for certain. Although female workers are at better levels both in terms of qualitative and quantitative sense in the tourism industry, it is a fact that female workers continue to face discrimination and obstacles in the sector. As female workers are fired in any little crisis or fluctuation in the tourism industry, more and more female workers become unemployed during the pandemic period (Kervankıran and Bağmancı, 2021: 267). There have been serious problems in the tourism industry during COVID-19 period.

Airline Transportation: Countries started to cancel flights as of the very first days within the scope of the precautions taken against COVID-19 virus and re-scheduling of flight plans and passengers' cancellation or postponement of their reservations took place (Şen and Bütün, 2021: 111). The aviation sector ensures economic growth, creates employment, and makes international trade and the tourism industry easier by having a single, fast transport network throughout the world. Working in a very active manner, the aviation sector was negatively affected as a result of travel limitations and gradual closure of borders by countries because of the global spread of the virus. Especially domestic and international flights were about to cease at a time related to tourism. One of the sectors that is affected by pandemic COVID-19 is the airline sector. Several airline companies reported losses in the pandemic period, and some of them went bankrupt (TURSAB, 2020).

There have been some arrangements to revive airline transportation within the pandemic COVID-19 period. Firstly, it was expressed notably to airline operators, plane operators, airline and airport staff, and ground services staff to guide on the recognition of indicators and symptoms of COVID-19 to increase awareness among aviation staff. Ground services staff and crew were informed on the precautions to prevent infection of COVID-19 such as social distance, hand hygiene, respiratory rules, refuse disposal, environment cleaning, and how to use (Şen and Bütün, 2021: 112).

Cruise Sector: Cruise sector is one of the travel segments that was affected by pandemic COVID-19 most. The sector, which achieved 42 billion dollars of growth before the pandemic, closed 2020 with a loss at a rate reaching 100% (TURSAB, 2020: 10). The sector is damaged in an economic sense, especially for 2020.

Congress, Meetings, and Entertainment Sector (MICE Sector): 87% of the organizers cancelled the organizations they would organize after the announcement of the pandemic, 66% of the events were pushed back. 70% of the face-to-face events were transferred to cyber platforms. 26 million workers working in this sector were negatively affected in occupational sense in COVID-19 period to global extent. 43% of the sector faced unemployment, reductions in salary, and other precautions. The loss that the sector suffered due to the cancelled or postponed trade fairs in the second quarter of 2020 reached up to 144.9 billion dollars. It is foreseen that the way that the sector does business will change radically in the future, and hybrid structures containing digital elements will be included in meetings and events in such a manner that the number of participants will increase (TURSAB, 2020: 10–11).

Accommodation Sector: One of the sectors that was intensely affected by the pandemic was the accommodation sector. Many hotels decreased their labour force significantly, and most of them had to close the enterprise permanently or temporarily. It is estimated that the demand on hotel enterprises will not reach pre- COVID-19 levels until 2023 (TURSAB, 2020: 11–12). Managers of accommodation enterprises are anxious about the increase in uncertainty during the pandemic period. Moreover, there are some problems with the incapability of paying the credits, not being able to make economic planning, a decrease in the number of qualified staff, a decrease in the facilities with high capacity, and the troubles brought about by complying with social distance rules in the facilities (Yüksek and Kalyoncu, 2021: 93).

Travel Agencies and Tour Operators: Being one of the most significant stakeholders in the tourism industry, travel agencies and tour operators are among the sectors that are most affected by COVID-19 process. The fact that massive and group travel came to stop during COVID-19 period entrapped the sector. The loss in holiday reservations was determined to be 85% compared to 2019. The loss the sector suffered at the end of 2020 was at a level of 73% (TURSAB, 2020: 12). During this period, travel agencies encountered some handicaps, such as a reduction in the motivation to travel, experienced staff's changing sector, decreasing motivation of personnel, changes in supply, cancellation of mass events, a decrease in interest in standard package tours, etc. (Yüksek and Kalyoncu, 2021: 91).

3.3 DISASTERS

Natural disasters such as earthquakes, tsunamis, whirlwinds, hurricanes, typhoons, floods, frost, droughts, volcanic eruptions, etc. caused serious influences on the lives of people throughout history. The destructive effect of a natural disaster shows itself directly in houses, hospitals, schools, factories, etc. buildings, where people spend their daily lives or meet their needs, and furthermore, reduces people's motivation for participation in economic life (Genç, 2018:87). Disasters affect tourism and the growth of countries in the long term (Ural, 2015:164). Since many tourism products are based on the internal assets of the local environment, tourism regions are generally close to areas that are subjected to natural threats such as severe winds volcanic activities. As tourism is a significant user of local infrastructure, including transportation networks, electric systems, and water supply systems, failure in the performance of these services may have negative results both in the short and long term for tourism, and this may lead to tarnishing the image of the destination (Byrd, 2007). Different touristic areas are more or less vulnerable to certain types of natural disasters compared to others (Faulkner, 2001).

3.3.1 DISASTERS AND EFFECTS ON TOURISM INDUSTRY

Tourism demand may fluctuate at different rates depending on the effect and growth of the crisis, and economic losses may become indispensable. The tsunami disaster that happened in 2004 in Asia affected the tourism and travel sectors in Malaysia, Indonesia, Sri Lanka, India, Thailand, Bangladesh, Maldives, Somalia, Tanzania, Kenya, and various countries in South Africa at different rates. By virtue of the fact that the countries have become dependent and bound to each other in the present world, any crisis that takes place on one side of the world is not exclusive to that country but may affect other regions and countries of the world. Touristic travel patterns are reformed as a result of this, and the tourism industry may encounter great problems and various negations and problems (Ritchie, 2004:670). Significant geological elements that affected tourism in recent years are as follows (COMCEC, 2017):

- The Indian Ocean tsunami in December 2004 affected 11 countries, notably Indonesia, Sri Lanka, India, and Thailand, most severely, and resulted in more than 230,000 deaths.
- Nepal earthquake that led to the deaths of approximately 9000 people in April 2015.
- The open sea earthquake and tsunami in March 2011 in the eastern part of Japan broke Fukushima Daiichi nuclear power station and resulted in the deaths of 22,000 people.
- The Katrina hurricane and great flood took place in 2010 in New Orleans, with more than 1800 loss of lives.
- Forest fires emerged as a result of human activities or a strike of lightning in natural areas; for instance, the forest fires happened in December 2016 around Gatlinburg, Tennessee, and resulted in the deaths of 14 people. A wide range of damage to touristic facilities or fires led to the deaths of more than 60 people in June 2017 in Portugal.

Some examples of the incidents occurred are presented in the following.

- Göçen et.al. (2011) stated in the research called crisis in Turkey and the effects of crisis on the tourism sector that the number of tourists decreased at a 24% rate compared to the previous year after the economic crisis in 1999 and August 17, 1999, Marmara earthquake.

- Natural disasters that strike touristic places in Indonesia frequently constitute a high level of uncertainty and threat (Kurniasari, 2017: 178). One of the natural disasters that threatens Indonesia most is volcanic eruption (Setyowati, Hadi and Ashari, 2013: 139). Another incident that happened in Indonesia was an earthquake experienced two times in a year on average (Sunarjo, Gunawan, and Pribadi, 2012: 2).
- Besides other incidents, in 2017, a set of hurricanes (Harvey, Irma, and Maria) are recorded in the Caribbeans, and a severe earthquake was recorded in Mexico, and these resulted in the highest loss recorded until now (135 billion US dollars) (Munich, 2018).
- From another point of view, Kim and Park (2016) concluded that radiation leakage arising from a nuclear accident is perceived by tourists as being subjected to a potential risk, and as a result, led to a serious decline in the number of people desiring to visit Japan (Estevão and Costa, 2020:6).
- 2000 of the loss of lives were tourists in six western provinces bearing up the effects of the tsunami that hit Thailand, where there were more than 5000 deaths (Smith and Henderson, 2008). As half of the deaths were tourists, it constituted a huge strike in Thailand economy. Local people, whose means of living depend on visitors' industries, list the reasons for the factors that lead to massive numbers of losses of lives as limited use of tsunami sirens, not having tsunami boards, inadequate knowledge of dangers among tourists, and limited evacuation training and responsibilities (Nguyen, Imamura, and Luchi, 2016: 4).
- Tourism destinations in Malaysia are indispensable for environmental disasters such as earthquakes, floods, and landslides. For instance, the earthquake took place in June 2015 damaged Kinabalu Park, a popular tourist attraction, and its neighbourhood in Malaysia's Sabah city seriously (Yee-Lee et al., 2017: 50).
- Avalanche disasters occur in the Alps. For example, four German skiers and a mountain guide died in the Austrian Oetztal Alps in April 1993; on May 1997, two skiers were buried while living in Pinzgau/Salzburg and in February 1999, a set of avalanches caused 39 losses of lives in the Paznaun valley (Hauke, 1999:13). 1998–99 year was one of the most difficult winters in terms of avalanches. Continuous snowfall and strong winds resulted in the accumulation of severe snow masses in the Alps. Apart from Austria, France was also affected by the avalanche, which caused 22 deaths in total.

- Huan, Beaman, and Shelby (2004) specified that foreign tourists felt that Taiwan was severely affected by 'September 21' earthquake after the 1999 earthquake and queries their security.

When tourists perceive a feeling of fear or high risk, an impression that can have sudden and unexpected negative impacts on a tourism destination may easily be created. Sönmez, Apostolopoulos, and Tarlow (1999) defended that tourism destinations will be subjected to negative influences such as tarnishing the image of tourism destinations, a lack of trust among potential travellers, and a sharp decline in incomes in the event of a lack of appropriate prevention actions. There have been various studies expressing the significance of disaster management planning (Hystad and Keller, 2005; Ritchie, 2004) to minimize the harm given to the tourism economy in case of a disaster (McCool, 2012). It may require recovery after disaster, reestablishment of touristic areas, and a slide in the image of tourism destinations (Ritchie, 2004).

3.3.2 *THE RECOVERY OF TOURISM INDUSTRY POST-DISASTER*

Natural disasters create long term effects on the lives of people, and as a result, the tourism industry can be affected negatively due to a decrease in the labour force, damage to tourism facilities, or tarnishing the image of a destination because of mismanagement. However, at the same time, tourist destinations can turn the natural disaster crisis into an opportunity for future investments in the region with well-organized crisis management and by increasing social support directed at the area. For that reason, as it is not possible to state when a natural disaster will take place exactly and to what extent it will be effective, disaster management becomes an important strategy for many tourist destinations (Genç, 2018).

Local or national authorities' study on various precautions to decrease the negative effects of natural disasters. For instance, supporting a biological shield through coastal plant cover has become a prevalent application in the past ten years as it provides natural protection against natural disasters such as hurricane waves or tsunamis and provides disaster management (Feagin et al., 2010: 1). However, none of these applications of precautions will work correctly without analysing the risks in the correct manner; therefore, risk analysis is conducted as the first step to eliminating the negative effects of natural disasters (Genç, 2018). Within this direction, creating a sense of security and attraction among the people following a disaster is quite critical to attracting the potential visitors to travel to the arrival area, guaranteeing them, and helping the affected region recover in an economic and functional sense (WTTC, 2018).

As pointed out by Laws and Prideaux (2005), a crisis affecting tourism will lead to a rapid decline in visitors entrance, and as a result of this, employment losses, a decrease in workload and profits, decreasing tax incomes for the government and overseas, and local entrepreneurs' not investing in these facilities. Unlike this opinion, Rittichainuwat (2008) expressed in his study that tourists will come to the tourism region even if those regions have the potential to be affected by disaster because tourists stated that knowing the effect of the disaster and the rescue work after the disaster motivates them.

It was underlined in the study of Tsai (2013:932) that disasters cannot be prevented with the current science and technology; however, the losses they can create and the effects of those on individuals, sectors, and governments can be decreased thanks to disaster management programs. For that reason, determination of how tourism disaster risk identification, assessment, control, and transfer will be performed through appropriate disaster risk management tools and the creation of a disaster risk and loss assessment model specific to the tourism industry are substantially significant and urgent tasks.

Here are some of the attempts of countries on crisis management.

- In 2003, there was a flood in Bukit Lawang, which is a tourism village in the North Sumatra province of Indonesia, and it caused the loss of more than 200 lives. The village was turned into a bunch of unplanned shops and guesthouses, expanding organically over a period of more than thirty years. Following this incidence, the government took steps to allow re-building in a more planned manner with additional stream management to take control of overflow in the future (COMCEC, 2017).
- While Ghaderi and others (2012) express that the tourism directed to Penang on Malaysia holiday island recovers 4–6 weeks after each crisis encountered, Carlsen and Hughes (2008) stated that the recovery of tourism from certain markets to the Maldives only lasts 3 months after a crisis. Rapid recovery in tsunami (even if it lasts longer in other markets) is considered to arise from effective campaigns, including familiarization trips for journalists, and increasing expenditures in big markets.
- Miller (2008) used an auto-ethnographic approach to analyze the improvement of tourism after disaster in New Orleans following Katrina Hurricane and expressed the brand strategy that defines New

Orleans as a 'turnaround city' to revive the local tourism industry. As well as creating a safe destination image, it is specified that sometimes disasters reveal new destination specifications and present the destination with an unprecedented recovery chance.

- Biran, Liu, Li, and Eichhorn (2014) examined the motivation of local tourists to visit Sichuan city of China after an earthquake. They found that natural disasters may lead to changes in the attractions of destinations, together with the fact that the commencement of a natural disaster and increasing interest in earthquakes by people resulted in an increase in visitors to Sichuan. Prayag (2016) called this phenomenon that generally emerges in the phase following disasters and surrounds the disaster theme, "post-disaster dark tourism". In the same way, Lin, Kelemen, and Kiyomiya (2017) searched blue tourism, which is a project on developing dark tourism following disasters in Japan after the 2011 Great East Japan Earthquake and Tsunami.
- It was because of the struggle with the effects of internal political issues and terrorist attacks in Nairobi and the coastal area that Kenya almost doubled its marketing budget in 2015/16 by creating incentive packages for charter-based tour programs and easing visa arrangements. A significant market for Southern African visitors.) These actions helped a recovery at a rate of approximately 16% in 2016 (Aviation, Travel, and Conservation News, 2017).
- Being an avalanche barrier and multi-functional building, 'Alpinarium' informs tourists about the avalanche calamities that happened in the past. The institution systematically collects data on natural mountain disasters and supports establishing networks with other institutions on disaster management (Peters and Pikkemaat, 2005). While this new building is an architectural and entertainment area for tourists, it also ensures security and safety for the center of Galtuer on the other hand.
- A non-profit organization, 'Alpine Safety and Information Centre' (ASI), was established to promote safety and security in mountain environments (Peters and Pikkemaat, 2005). Its purpose is to provide a communication bridge for all participant organizations and local institutions. Furthermore, a general discussion on ensuring the safety and security of tourists was commenced in the Alps winter sports center in Austria. The ESIS internet platform was developed with a communication and content management system after disasters to prevent this kind of non-preparation in the future.

In conclusion, crisis that slow down or hinder the improvement of the tourism and travel sector or affect the general course of it in a negative manner disturb tourists and may direct them to safer countries by dissuading them from travelling to countries where there is not safety and security available and where the risk is high (Olalı and Timur, 1988) and without freedom of mobility despite their intense desires. Likewise, tourists crave safe harbours. People can change their holiday plans and choices or cancel them even if they have already made them or they can be inclined to safer destinations when there is not safety and security (Köşker, 2017). When a destination encounters an incident/event that affects market perceptions negatively, regaining trade requires additional efforts from destination authorities and the tourism industry. Some of them are summarized above; however, fundamental elements are familiarization trips and other advertisement efforts, state commitment and leadership, reduction of local costs and taxes, and ensuring subsidies.

3.4 CONCLUSIONS

Various epidemics and natural disasters have been encountered throughout the history of humankind at various times. Particularly, it was assumed that epidemics would disappear or the effects of it would be weak to a large extent thanks to modern medicine. However, the never-ending desires of human beings, wasting resources unnecessarily, and excessive destruction of nature, the environment, and air pollution have resulted in increase in epidemics and natural disasters, contrary to the expectation of their disappearance. Furthermore, studies on epidemics and natural disasters foresee that natural disasters will be faced more, and more and new virus epidemics will enter our lives within shorter periods of time and lead to more negative conclusions.

Although epidemic SARS is less lethal compared to MERS (MERS with a 34.4% moratality rate and SARS with a 9.6% mortality rate), its effects on the global economy is between 30 and 100 billion dollars as epidemic SARS spreads to wider geographies. Even though pandemic COVID-19 is less fatal than SARS (1–3% until now), it spread worldwide. Despite the greatest limitation efforts and meticulous precautions of many countries in the world in relation to COVID-19 compared to the others in history, taking the pandemic under control could not be achieved (Khalilzadeh, 2020:3). From this point, more people were affected by the economic effects of diseases spreading to wider areas. Similarly, pandemic COVID-19 is also a disease spread to the world and together with its effects associated with health, it has affected several sectors and people in economic sense.

It is indispensable that there have been changes in tourists' attitudes and travel choices during the pandemics/epidemics happening in the world. There are also changes in the attitudes of local people in destination areas not just in the attitudes and travel choices of tourists.

It was determined that there was an increase in people's participation in leisure time and recreation activities after the SARS epidemic in 2004 compared to the pre-epidemic period. The interests of people in outdoor recreation and tourism activities increased to spend time with their acquaintances and to keep away from the effects of stressed days concerning their health (Marafa and Tung, 2004). Similarly, there have been changes in tourist activities within the course of the COVID-19 epidemic. To give an example, it has been observed that the interest in health tourism, the demand for nature-based tourism activities (trekking and camping etc.), and participation in individual tourism activities have increased with the pandemic under control and the acceleration of vaccination works. Quieter and nature-based tourism destinations are preferred rather than extremely crowded destinations with intense mass tourism. Contagious diseases, quarantine status, safe travel, and cleaning and hygiene factors come into prominence when choosing destinations compared to the pre-pandemic period. Moreover, the attitude towards travelling with individual vehicles as the safest way of travelling rather than using mass transportation vehicles has increased.

It is expected that the sector will recover until 2023 following COVID-19. However, it is foreseen that there will be structural and functional changes in the sector. For example, it is expected that the tendency towards localization will increase, and young labour force will be employed, and different destinations will be preferred over current destinations. It is assumed that international travel activities will continue along with the increase in the tendency towards local and sustainable tourism varieties, and destinations will concentrate on sustainability issues. It is required that countries formulate sustainable improvement strategies and development plans for the tourism sector. State administrations shall define incentive packages and policies during and after pandemics to ensure recovery of the sector. Enterprises shall pay more attention to health, safety, and hygiene issues as the cognitive attitudes of tourists during travel changes after pandemics. Travel agencies shall prepare smaller group activities and private tours following pandemics. It is estimated that interest in eco-tourism tours and health tourism package programs will increase due to the pandemic. It is expected that interest in natural accommodation facilities will increase due to the pandemic.

Intermediary phases of crisis management will turn into longer term precautions gradually. These are included under the following titles:

- Improvement of infrastructure
- Investment climate
- Improvement of human resources
- Re-creation and re-position of the image
- Market and product variety
- Familiarization trips and increasing advertisement
- State commitment and leadership
- Cost reduction

KEYWORDS

- **pandemic**
- **disaster**
- **tourism industry**
- **the recovery of tourism industry**

REFERENCES

Alaeddinoğlu, F.; Rol, S. COVID-19 Pandemisi ve Turizm Üzerindeki Etkileri. *Van Yüzüncü Yıl üniversitesi Sosyal Bilimler Enstitüsü Dergisi*, Salgın Hastalıklar Özel Sayısı 2020; pp 233–258.

Aslan, R. Tarihten Günümüze Epidemiler, Pandemiler ve COVID-19. *Göller Bölgesi Aylık Ekonomi ve Kültür Dergisi* **2020,** *8* (85), 35–41.

Au, A. K.; Ramasamy, B.; Yeung, M. C. The Effects of SARS on the Hong Kong Tourism Industry: An Empirical Evaluation. *Asia Pac. J. Tour. Res.* **2005,** *10* (1), 85–95.

Aviation, Conservation and Travel News, Kenya's Tourism Arrivals Grew by Over 16 Percent in 2016. https://wolfganghthome.wordpress.com/2017/02/01/kenyas-tourism-arrivals-grew-by-over-16-percent-in-2016/ (Accessed 10 Aug 2021).

Aydın, B.; Doğan, M. Yeni Koronavirüs (COVID-19) Pandemisinin Turistik Tüketici Davranışları ve Türkiye Turizmi Üzerindeki Etkilerinin Değerlendirilmesi, *Pazarlama Teorisi ve Uygulamaları Dergisi* **2020,** *6* (1), 93–115.

Baum, T.; Hai, N. T. T. Hospitality, Tourism, Human Rights and the Impact of COVID-19. *Int. J. Contemp. Hosp. Manage.* **2020,** *32* (7), 2397–2407.

Benjamin, S.; Dillette, A.; Alderman, D. H. "We Can't Return to Normal": Committing to Tourism Equity in The Postpandemic Age. *Tour. Geogr.* **2020,** *22* (3), 476–483.

Biran, A.; Liu, W.; Li, G.; Eichhorn, V. Consuming Post-Disaster Destinations: The Case of Sichuan, China. *Ann. Tour. Res.* **2014,** *47*, 1–17.

Boone, L.; Haugh, D.; Pain N.; Salins, V. Tackling the Fallout from COVID-19. In *Economics in the Time of COVID-19*; Baldwin, R.; di Mauro, B. W., Eds.; CEPR Press: London, 2020.

Byrd, E. T. Stakeholders in Sustainable Tourism Development and Their Roles: Applying Stakeholder Theory to Sustainable Tourism Development. *Tour. Rev.* **2007,** *62* (2), 6–13.

Carlsen, J. C.; Hughes, M. Tourism Market Recovery in the Maldives After the 2004 Indian Ocean Tsunami. *J. Travel Tour. Market.* **2008,** *23* (2–4), 138–149.

COMCEC. *Risk & Crisis Management in Tourism Sector: Recovery from Crisis in the OIC Member Countries*; Turkey, 2017.

Çeti, B.; Ünlüönen, K. Salgın Hastalıklar Sebebiyle Oluşan Krizlerin Turizm Sektörü Üzerindeki Etkisinin Değerlendirilmesi. *AHBVÜ Turizm Fakültesi Dergisi* **2019,** *22* (2), 109–128.

Çetin, G.; Göktepe, S. COVID-19 Pandemisinin Turizm Endüstrisi Üzerindeki Etkileri. In *COVID-19 Pandemisinin Ekonomik, Toplumsal ve Siyasal Etkileri*; Demirbaş, D., Bozkurt, V., Yorgun, S., Eds.; İstanbul Üniversitesi Yayınevi: İstanbul, 2020; pp 87–97.

Estevão, C.; Costa, C. Natural Disaster Management in Tourist Destinations: A Systematic Literature Review. *Eur. J. Tour. Res.* **2020,** *25*, 2502.

Faulkner, B. Towards a Framework for Tourism Disaster Management. *Tour. Manage.* **2001,** *22*, 135–147.

Feagin, R. A.; Mukherjee, N.; Shanker, K.; Baird, A. H.; Cinner, J.; Kerr, A. M.; Koedam, N.; Sridhar, A.; Arthur, R.; Jayatissa, L. P.; Lo Seen, D.; Menon, M.; Rodriguez, S.; Shamsuddoha, Md.; Dahdouh-Guebas, F. Shelter from the Storm? Use and Misuse of Coastal Vegetation Bioshields for Managing Natural Disasters. *Conserv. Lett.* **2010,** *3* (1), 1–11.

Genç, R. Catastrophe of Environment: The Impact of Natural Disasters on Tourism Industry. *J. Tour. Advent.* **2018,** *1* (1), 86–94.

Ghaderi, Z.; Mat Som, A. P.; Henderson, J. Case Study: Tourism Crises and Island Destinations: Experiences in Penang, Malaysia. *Tour. Manage. Perspect.* **2012,** *2* (3), 79–84.

Göçen, S.; Yirik, Ş.; Yılmaz, Y. Türkiye'de Krizler ve Krizlerin Turizm Sektörüne Etkileri. *Süleyman Demirel Üniversitesi İktisadi ve İdari Bilimler Dergisi* **2011,** *16* (2), 493–509.

Grubman, M.; Baxt, B. Foot-and-Mouth Disease. *Clin. Microbiol. Rev.* **2004,** *17* (2), 465–493.

Gu, H.; Wall, G; The Effects of SARS on China's Tourism Enterprises. *Turizam: međunarodni znanstveno-stručni časopis* **2006,** *54* (3), 225–234.

Haque, T. H.; Haque, O. The Swine Flu and its Impacts on Tourism in Brunei. *J. Hosp. Tour. Manage.* **2018,** *36*, 92–101.

Hauke, B. *Die weiße Gefahr*; ERC Fankona Rückversicherungs-AG: München, 1999.

Huan, T. C.; Beaman, J.; Shelby, L. No-Escape Natural Disaster: Mitigating Impacts on Tourism. *Ann. Tour. Res.* **2004,** *31*, 255–273.

Hystad, P.; Keller, P. Disaster Management: Kelowna Tourism Industry's Preparedness, Impact and Response to a 2003 Major Forest Fire. *J. Hosp. Tour. Manage.* **2005,** *13*, 44–58.

Joo, H.; Maskery, B. A.; Berro, A. D.; Rotz, L. D.; Lee, Y. K.; Brown, C. M. Economic Impact of the 2015 MERS Outbreak on the Republic of Korea's Tourism-Related Industries. *Health Secur.* **2019,** *17* (2), 100–108.

Karadeniz, E.; Beyaz, F. S.; Ünlübulduk, S. N.; Kayhan, E. COVID-19 Salgınının Turizm Sektörüne Etkilerinin ve Uygulanan Stratejilerin Değerlendirilmesi: Otel Yöneticileri Üzerinde Bir Araştırma. *Türk Turizm Araştırmaları Dergisi* **2020,** *4* (4), 3116–3136.

Karakaş, M. COVID-19 Salgınının Çok Boyutlu Sosyolojisi ve Yeni Normal Meselesi. *İstanbul Üniversitesi Sosyoloji Dergisi* **2020,** *40* (1), 541–573.

Karcıoğlu, Ö. COVID-19: Epidemiyolojik Bilgilerimiz ve Hastalığın Dünyadaki Gidişi. *J. ADEM* **2020,** *1* (1), 55–71.

Kervankıran, İ.; Bağmacı, M. F. COVID-19 Sürecinde Türkiye Turizminin Mekansal Görünümü: Hangi İl Nasıl Etkilendi? *Coğrafi Bilimler Dergisi/Turk. J. Geogr. Sci.* **2021,** *19* (1), 263–287.

Khalilzadeh, J. The Future of The Global Tourism System Post COVID-19, 2020. https://business.ecu.edu/wp-content/pv-uploads/sites/152/2020/04/The-Future-Of-The-Global-Tourısm-System-Post-COVID-19Final.pdf (Accessed 6 June 2021).

Kim, J. S.; Park, S. H.A. Study of the Negotiation Factors for Korean Tourists Visiting Japan Since the Fukushima Nuclear Accident Using Q-methodology. *J. Travel Tour. Market.* **2016,** *33* (5), 770–782. DOI: 10.1080/10548408.2016.1167395.

Kiper, V. O.; Saraç, Ö.; Çolak, O.; Batman, O. COVID-19 Salgınıyla Oluşan Krizlerin Turizm Faaliyetleri Üzerindeki Etkilerinin Turizm Akademisyenleri Tarafından Değerlendirilmesi. *Balıkesir Üniversitesi Sosyal Bilimler Enstitüsü Dergisi* **2020,** *23* (43), 527–551.

Köşker, H. Krizlerin Turizm Sektörüne Etkileri Üzerine Bir Araştırma: 2016 Yılı Türkiye Örneği. *Uluslararası Hakemli Sosyal Bilimler E-Dergisi* **2017,** 62.

Kurniasari, N. Strategi Penanganan Krisis Kepariwisataan Dalam Kebijakan Badan Nasional Penanggulangan Bencana (BNPB). *Media Tor* **2017,** *10* (2), 177–189.

Laws, E.; Prideaux, B. Crisis Management: A Suggested Typology. *J. Travel Tour. Market.* **2005,** *19* (2–3), 1–8.

Lin, Y.; Kelemen, M.; Kiyomiya, T. The Role of Community Leadership in Disaster Recovery Projects: Tsunami Lessons from Japan. *Int. J. Project Manage.* **2017,** *35* (5), 913–924.

Marafa, L. M.; Tung, F. Changes in Participation in Leisure and Outdoor Recreation Activities Among Hong Kong People During the SARS Outbreak. *World Leisure J.* **2004,** *46* (2), 38–47.

Mason, P.; Grabowski, P.; Du, W. Severe Acute Respiratory Syndrome, Tourism and the Media. *Int. J. Tour. Res.* **2005,** *7* (1), 11–21.

McCool, B. N. The Need to Be Prepared: Disaster Management in the Hospitality Industry. *J. Busi, Hotel Manage.* **2012,** *1*, 1–5.

Miller, S. D. Disaster Tourism and Disaster Landscape Attractions After Hurricane Katrina: An Auto-Ethnographic Journey. *Int. J. Culture Tour. Hosp. Res.* **2008,** *2* (2), 115–131.

Mizrachi, I.; Fuchs, G. Should We Cancel? An Examination of Risk Handling in Travel Social Media Before Visiting Ebola-Free Destinations. *J. Hosp. Tour. Manage.* **2016,** *28*, 59–65.

Munich, R. Press Release: Natural Catastrophe Review: Series of Hurricanes Makes 2017 Year of Highest Insured Losses Ever, 2018. https://www.munichre.com/en/media-relations/publications/press-releases/2018/2018–01–04-press-release/index.html (Accessed 26 Sept 2021).

Novelli, M.; Burgess, L. G.; Jones, A.; Ritchie, B. W. 'No Ebola…Still Doomed'—The Ebola-Induced Tourism Crisis. *Ann. Tour. Res.* **2018,** *70*, 76–87.

Nguyen, D. N.; Imamura, F.; Kanako Luchi, K. Disaster Management in Coastal Tourism Destinations: The Case for Transactive Planning and Social Learning. *IRSPSD Int.* **2016,** *4* (2), 3–17.

Olalı, H.; Timur, H. *Turizm Ekonomisi*; Ofis Ticaret Matbaacılık Şti.: İzmir, 1988.

Page, S.; Yeoman, I.; Munro, C.; Connell, J.; Walker, L. A Case Study of Best Practice—Visit Scotland's Prepared Response to an Influenza Pandemic. *Tour. Manage.* **2006,** *27* (3), 361–393.

Pavli, A.; Tsiodras, S.; Maltezou, H. C. Middle East Respiratory Syndrome Coronavirus (MERS-Cov): Prevention in Travelers. *Travel Med. Infect. Dis.* **2014,** *12* (6), 602–608.

Peters, M.; Pikkemaat, B. Crisis Management in Alpine Winter Sports Resorts-the 1999 Avalanche Disaster in Tyrol. *J. Travel Tour. Market.* **2005,** *19* (3/4), 9–21.

Prayag, G. It's Not All Dark!: Christchurch Residents' Emotions and Coping Strategies with Dark Tourism Sites. In *Business and Post-Disaster Management*; Hall, C. M., Malinen, S., Vosslamber, R., Wordsworth, R., Eds.; Routledge: Oxford, 2016; pp 155–166.

Ritchie, B. W. Chaos, Crisis and Disaster: A Strategic Approach to Crisis Management in The Tousim Industry. *Tour. Manag.* **2004,** *25*, 669–683.

Rittichainuwat, B. N. Responding to Disaster: Thai and Scandinavian Tourists' Motivation to Visit Phuket, Thailand. *J. Travel Res.* **2008,** *46* (4), 422–432.

Setyowati, S.; Hadi, B. S.; Ashari, A. Pengembangan Sistem Informasi Bahaya Erupsi Untuk Pengelolaan Kebencanaan Di Lereng Selatan Gunung Merapi. *Majalah Geografi Indoensia* **2013,** *7* (2), 138–148.

Smith, R.; Henderson, J. C. Integrated Beach Resorts, Informal Tourism Commerce and The 2004 Tsunami: Laguna Phuket in Thailand. *Int. J. Tour. Res.* **2008,** *10* (3), 271–282. DOI: 10.1002/jtr.659.

Sönmez, S.; Apostopoulos, Y.; Tarlow, P. Tourism in Crisis: Managing the Effects of Terrorism. *J. Tour. Res.* **1999,** *38* (1), 13–18.

Sunarjo, Gunawan, M. T.; Pribadi, S. *Gempa bumi*; Badan Meteorologi Klimatologi dan Geofisika: Jakarta, 2012.

Şen, G.; Bütün, E. COVID-19 Salgınının Havacılık Sektörüne Etkisi: Gig Ekonomisi Alternatifi. *J. Aviation Res.* **2021,** *3* (1), 106–127.

Tsai, C-H. Multi-Hazard Risk Assessment and Management in Tourism Industry- A Case Study from the Island of Taiwan. *Int. J. Soc. Manage. Econ. Busi. Eng.* ***2013,*** *7* (8).

TURSAB, COVID-19 Sürecinde Türkiye ve Dünya Turizmi Değerlendirmesi https://www.tursab.org.tr/duyurular/COVID-19-surecinde-turkiye-ve-dunya-turizmi-degerlendirmesi (Accessed 15 Aug 2020).

UNWTO. Impact Assessment of the COVID-19 Outbreak on International Tourism. https://www.unwto.org/impact-assessment-of-the-COVID-19-outbreak-on-international-tourism (Accessed 28 Sept 2021).

Ural, M. Turizmin Sürdürülebilirliği için Risk Yönetiminin Önemi. *Balıkesir Univ. J. Soc. Sci. Instit.* **2015,** *18* (3).

Wang, Y. The Impact of Crisis Events and Macroeconomic Activity on Taiwan's International Inbound Tourism Demand. *Tour. Manage.* **2009,** *30* (1), 75–82.

Wilder-Smith, A. The Severe Acute Respiratory Syndrome: Impact on Travel and Tourism. *Travel Med. Infect. Dis.* **2006,** *4* (2), 53–60.

World Travel and Tourism Council (WTTC). Caribbean Resilience and Recovery: Minimising the Impact of the 2017 Hurricane Season on the Caribbean's Tourism Sector. London. https://www.wttc.org/-/media/files/reports/2018/impact-of-the-2017-hurricane-season-on-the-caribbean-spread.pdf (Accessed 20 Aug 2021).

World Travel and Tourism Council (WTTC). Travel & Tourism Economic Impact from COVID-19. https://wttc.org/Research/Economic-Impact (Accessed 20 Sept 2021).

Yee-Lee, C.; Lin-Sea, L.; Kwang-Jing, Y. Natural and Consequence of Tourism Industry's Susceptibility to Natural Disaster—A Conceptual Framework Using Input-Output Model. *Josai Contemp. Policy Res.* **2017,** *10* (1).

Yüksek, G.; Kalyoncu, M. COVID-19 küresel Salgınının Turizm Sektörü Üzerindeki Etkileri, *J. Gastronomy Hosp. Travel* **2021,** *4* (1), 85–101.

Zeng, B.; Carter, R. W.; De Lacy, T.Short-Term Perturbations and Tourism Effects: The Case of SARS in China. *Curr. Iss. Tour.* **2005,** *8* (4), 306–322.

CHAPTER 4

Redesigning, Reinventing, and Rebuilding Tourism Activities

KÜBRA GÜNDOĞDU[1], ANUKRATI SHARMA[2], and GÜL ERKOL BAYRAM[3]

[1] *Institution of Social Sciences, Tourism Management, Selçuk University, Konya, Turkey*

[2]*Department of Commerce and Management, University of Kota, Kota, India*

[3] *Department of Guiding, School of Tourism and Hospitality, University of Sinop, Turkey*

ABSTRACT

Tourism is a sector that has significant economic development power all over the world. This situation is seen as a serious competitive advantage in tourism all over the world (Antara and Sumarniasih, 2017: 35). The COVID-19 pandemic, which emerged in the last 2 years, showed us the economic power of tourism. However, this demonstration took place with the collapse of the tourism sector, which has a weak structure. This study deals with the redesign and rediscovery of the traditional structure of the tourism sector and the reconstruction of tourism for destinations in accordance with societies, economic and social structures, cultures, and natural life (Chang et al., 2020). The study was handled conceptually and was created with the document analysis technique, one of the qualitative research methods. As a result of the study, it has been understood that the tourism sector has come to a standstill during the crisis processes of tourism._At this point, it was

Dynamics of the Tourism Industry: Post-Pandemic and Post-Disaster Perspectives and Strategies.
Gül Erkol Bayram and Anukrati Sharma (Eds.)

concluded that tourism should be restructured under the umbrella of sustainability, and different types of tourism should be developed.

4.1 INTRODUCTION

While tourism has become an indispensable source of income for the whole world, its uncontrolled growth has had serious consequences. Developing tourism uniformly in every destination with the effort of generating only economic income and ignoring the natural environment, the social and social structure in this process brought irreversible deterioration in these structures over time. Tourism itself has been affected by these deteriorations and has begun to lose its resources. The efficient use of scarce resources and their transfer to future generations are of great importance not only for life but also for the sustainability of tourism (Saarinen, 2009). In addition, it has been understood that the traditional tourism structure is weak as a result of the global pandemic of COVID-19, which emerged in China at the end of 2019 and covered the whole world in a short time (Sheresheva, 2020). For this reason, as it should be in all areas, it should be restructured in accordance with the society, environment, and wildlife with the understanding of sustainability in tourism (Chang et al., 2020). In recent years, the fact that individuals act with the awareness of sustainability, as well as seek travel for their personal interests, has created an important opportunity for the redesign of tourism. In particular, the understanding of benevolent tourism has a significant impact on the development of underdeveloped and developing countries with the support of tourism. The charitable activities and donations of tourists from the USA, Canada, England, and Australia to protect wildlife in Kenya have greatly affected the region (Barnes and Eagles, 2004: 35). Another issue that has attracted attention in the last 2 years, when tourism activities have been eliminated, is how much recreational activities are needed. Recreation activities allow people to evaluate their spare time outside of their daily routines. Tourism has come to a standstill, and the importance of recreational activities should be taken into account in this process (Sun and Walsh, 1998: 325; Fang et al., 2021).

This study is about redesigning, reinventing, and rebuilding tourism activities according to the changing consumption habits in recent years. Within the scope of this subject, it has been researched how the consumption habits in tourism have changed today, and it has been tried to determine what is effective in the reorganization of tourism activities. In the research, it has been understood that the understanding of sustainability has started to play an active role in today's tourism consumption habits and that special interest tourism types have emerged under the umbrella of sustainability. In addition, it has been understood that it has started to gain importance in recreational activities in recent years.

4.2 REDESIGNING IN TOURISM ACTIVITIES

Tourism event has occurred for many different reasons throughout history. In the early ages, the reasons for travel were for basic needs such as food and shelter. In the Middle Ages, travels began to take place for reasons such as religious, political, adventure, and curiosity. Before the industrial revolution, travel was seen as an event that only upper-class people could afford. The technological developments and the collapse of the feudal structure in this period are revolutionary in tourism. In the process that started with the industrial revolution, tourism activities came to the fore, and big tours started to be organized under the name of grand tours. In the early ages, travels on foot or by riding animals took a long time, but with the development of vehicles, travel times were shortened significantly (Kozak et al., 2013). Labor law studies, the abolition of international travel restrictions, and the right to paid leave for employees have led people to travel. Workers' law studies, the abolition of international travel restrictions, and the right to take paid leave to employees have led people to travel. Tourism activities that developed in this process were focused on sea, sand, sun, and silence. The fact that people travel to similar regions in masses and that the host regions see tourism as a short-term gain has brought many problems. In the 21st century, along with the rapidly increasing population, many social, cultural, environmental, and similar problems have been encountered. Along with these problems, some people have started to turn to different types of tourism and activities to relax and have fun, especially in destinations that attract great attention. Types of tourism, which differ according to personal experience and preferences, entered the tourism literature as alternative tourism and later started to be referred to as special interest tourism (Çelik, 2018).

At the end of the 20th century, one of the important issues discussed in the world with the problems brought on by globalization has been sustainability. The concept of sustainable development emerged in 1987 with the report "Our Common Future" published by the World Commission on Environment and Development. Focusing on sustainable development and the continuity of economic, social, and natural environmental resources, experts emphasized the problem of exceeding the carrying capacity in these dimensions.

Sustainable development has also affected the tourism sector, which is one of the fastest-growing sectors in the world, as it has in all sectors. Exceeding the carrying capacities in the destinations of interest has had an effective effect on the deterioration of the natural, historical, and cultural structure. In particular, the inability to provide a fair income distribution in the host societies and the regional and seasonal effects of tourism has caused tourism and tourists to be viewed with negative thoughts. These events in the 1980s and the uncontrolled development of mass tourism have been major factors in the formation of the concept of sustainable tourism.

Considering the effects of tourism activities, it has been concluded that there are many benefits and harms. Considering the continuity of tourism resources and tourism-related resources in the last 50 years, it has been concluded that tourism should be redesigned. It is seen that the uncontrolled development of tourism activities causes serious damage before tourism can reach the marginal benefit. This lack of control causes serious damage to the economy as well as to the historical structures, cultural areas, society, and natural environment in destinations (Wijaya, 2019: 66). At the point of redesigning tourism activities, sustainable, responsible, and community-based ecological tourism approaches emerge. As in all areas, there is a need for efficient use of scarce resources in the field of tourism. The fact that tourism has a structure that is most affected and most influential from other sectors also makes this issue more important (Hughes and Scheyvens, 2016; Paskova and Zelenka, 2018: 542).

The COVID-19 pandemic, which emerged in 2019, showed us how fragile tourism and the studies in the field of tourism are. The fact that tourism destinations have been built for only one tourism area has revealed that it has insufficient power in times of crisis. The inadequacy of the uniform design of tourism activities in the face of the crises experienced in the last two years has revealed the importance of recreational activities. The fact that the travel and activities preferred by local and foreign tourists who stay away from crowded tourist areas during the pandemic process are rural tourism and recreational activities emphasized the importance of this situation (Özçoban, 2020; Üner, 2021).

4.3 REINVENTING IN TOURISM ACTIVITIES

With the pandemic crisis that took place in the last two years, the whole world has realized that the tourism and entertainment sectors are built on an unstable foundation. The COVID-19 global pandemic has caused leisure and domestic tourism revenues to drop by more than 50% to $2.86 Trillion (Abbas et al., 2021). It is thought that considering the tourism capacity of each destination differently, creating service areas suitable for those destinations, and developing tourism activities in accordance with the cultural, natural, and social structure of the destination will increase the durability of tourism. In addition, it has been observed that the use of technology in tourism is insufficient (Scott et al., 2008: 11).

In this process, where the power of tourism to resist the crisis was insufficient, one of the situations that was noticed was the insufficient value given to rural tourism. While the importance of individual travels increased during the pandemic process, these travels mostly increased the interest in rural areas, nature, and wildlife. During the crisis, when mass tourism was out of the way for the tourism sector, people understood the importance of nature and wildlife better. The development of natural life, societies, and historical and cultural resources by preserving and achieving economic efficiency in tourism will be possible with the understanding of sustainability, responsibility, and tourism alternatives developing accordingly (Arslan and Kendir, 2020: 3668–3671).

4.3.1 *SUSTAINABLE AND RESPONSIBLE TOURISM*

Sustainable tourism is a concept that gathers all types of tourism under the approach of sustainability rather than being a type of tourism. In this context, sustainable tourism has adopted the principle of realizing all kinds of tourism in line with basic objectives such as increasing the quality of life of the local people, protecting the rights of future generations, improving the natural environment and wildlife standards, the protection of life, social structure and cultural values, the vital activities of persons with disabilities, and the equal provision of their travel rights. This situation necessitated the reorganization of tourism activities. Along with the development of sustainability awareness, the increase in the level of education in recent years and the development of conscious production and consumption habits people have also been reflected in their travels.

The United Nations World Tourism Organization has listed what needs to be done within the scope of sustainable tourism as follows (UNWTO, 2021):

- To encourage the efficient use of environmental resources, one of the main resources of tourism, by ensuring the continuity of the natural environment and biological diversity without hindering the development of technology.

- Respecting the lives and cultures of the host communities and preserving their traditions, customs, and traditions with the understanding of cultural tolerance in tourism.
- To seize employment opportunities in tourism, to ensure fair distribution of tourism revenues for host communities, and to develop long-term economic plans.

Especially in the last 20 years, the increase in mechanization and intense and stressful city life have led people to rural life and natural environmental activities. The basis of tourism activities developed around sustainability is tourism activities that come together with constructive works such as the protection and structuring of natural cultural life and historical assets, the development of host societies, the provision of economic and social equality, and the provision of economic and social equality. They are activities where tourists can realize their personal desires (Giampiccoli and Saayman, 2017). This attitude of tourism consumers has passed into the tourism literature as responsible tourism. Responsible tourism includes issues related to employee rights in the sector, development of local communities, respect for gender equality and the rights of future generations, ensuring justice in trade and employment of women and children in the organization, and implementation of its activities. One of the important points of responsible tourism is that it aims to minimize the negative effects arising from cultural and social differences and consumption habits between tourists and host societies. Within the scope of the Cape Town Declaration, the following views were presented on the benefits of responsible tourism in tourism (Frey and George, 2010):

- Responsible tourism supports the fair distribution of income in tourism and the development of the local economy;
- Improves the well-being of host communities;
- It strengthens the relationship between local people and tourists and affects the establishment of an effective bond between the two parties;
- Aims at intercultural respect and tolerance;
- Facilitates the access of physically disabled individuals to tourism products.

Mathew and Sreejesh (2017) in their study of the effects of responsible tourism on a sample of 432 people in three different destinations in India concluded that there is a positive relationship between responsible tourism activities and local communities' perspectives on tourism. It has been stated that the reason for this is that responsible tourism activities have positive effects on the quality of life in local communities. Spenceley et al. (2002), one of the most important benefits of responsible tourism is that it reveals new entrepreneurship opportunities. The business opportunities that develop around responsible tourism activities provide entrepreneurial opportunities to local communities. In recent years, where environmentalist consumption habits have gained importance, ecotourism, community-based tourism, agro–tourism, voluntary tourism, and smart and accessible tourism, such as responsible tourism, play an important role in shaping and redesigning tourism activities.

4.4 REBUILDING IN TOURISM ACTIVITIES

Man has been in search of innovation throughout his life and people constantly sought alternatives to the service and product they received and wanted more. In the last 50 years, this search has been felt intensely in tourism, and this search has been tried to be answered. Sports, art, science, culture, wildlife and more have appeared as alternatives in tourism. This change of tourism, which is a strong source of economy, has become an opportunity for countries (Cohen, 1987: 13–15). In the past, the activities carried out only for mass tourism have only failed the efforts of countries to spread tourism to the whole year and the whole area. Today, the diversification of tourism and its development in this direction have helped the economic and social development of the destinations, while increasing the reasons for travel and the satisfaction level of the visitors (Briedenhan, 2009: 98–100).

The change in the structure of tourism activities has affected the local communities as well as the visitors. In classical tourism management, the host societies see the incoming visitors as invaders, which is based on the creation of tourism activities without considering the local society and environment. For this reason, ensuring the participation of local people and local governments in the process of creating the tourism structure affects the efficiency and satisfaction level of tourism. It is thought that the local communities that know the destinations best will be the best guides for the incoming tourists. In addition, the host societies' view of incoming visitors as invaders is based on the creation of tourism activities without considering the local society and environment (Güneş, 2011: 55).

Increasing interest in travel types that develop according to their special interests and adopting the understanding of sustainability in tourism have supported the formation of benevolent, environmentalist, protective, educational, and experience-based tourism activities in tourism activities. As the awareness level of the visitors increased, the social, economic, and environmental benefits of the trips increased in the same direction. This approach to tourism has led to the development of new types of tourism that will increase the gains of visitors, host communities, and natural life (Adabalı, 2021: 204). Today, many trips are organized to benefit societies and the environment (Aykın, 2018). In recent years, ecotourism, community-based tourism, agro–tourism, voluntary tourism, and accessible tourism are at the forefront of the travels carried out and discovered for this purpose.

4.4.1 ECOTOURISM

Ecotourism aims to harmonize all aspects of tourism activities with natural and social life. Güneş (2018) stated that the most important point in the development of ecotourism is minimizing the negative effects caused by the activities of tourism. Consumption habits made with a sense of responsibility toward nature, society, and wildlife have also increased the interest in ecotourism. In the study by Blamey (2001), it is suggested that the critical point in the development of ecotourism is the product–market relationship and stated that the tourism product that is not suitable for the market will prevent the development of ecotourism activities. It has been observed that the level of welfare has increased in destinations where ecotourism activities are developed. Arranging tourism activities in accordance with the structure of destinations has increased the number of studies carried out by local communities to enrich tourism activities (Butarbutar and Soemarno, 2013). One of the leading problems of tourism is the design of tourism enterprises in a way that negatively affects the structure of the regions. One of the subjects of sustainability and ecotourism is designing the structure of accommodation, food and beverage, and other touristic businesses that are not suitable for the structure of the regions and cause the deterioration of the natural structure, in accordance with the destinations. It is very important that the businesses where tourism activities are carried out are suitable for the social and natural structure (Jaafar and Maideen, 2012).

4.4.2 COMMUNITY-BASED TOURISM

Community-based tourism aims to solve the problems caused by the uncontrolled tourism activities of underdeveloped and developing countries and to eliminate poverty with the fair distribution of tourism revenues (Aydemir and Kazoğlu, 2016). In community-based tourism, local communities can participate in all phases of tourism. Community-based tourism, especially in underdeveloped countries and destinations, provides the traditional and basic livelihoods of the poor people to support their activities and play an active role in tourism (Karacaoğlu et al., 2016). Okazaki (2008), In the study community-based tourism under the theories of participation of local people, redistribution of power, and creation of social capital. In the study is stated that the first step in this process is to determine the current situation of the society. Afterward, it was suggested to determine incentive methods for tourism suitable for different societies. The inclusion of local communities in tourism activities made it necessary to organize tourism activities according to the structure of societies. Köroğlu and Karaman (2014) stated in their study that regional cooperatives play an important role in tourism management and listed the qualifications to be followed within the scope of community-based tourism as follows:

- One of the serious problems in tourism is the situation of injustice in income distribution. While the local communities and the natural structure they live in are the most affected by tourism activities, the inability to gain economic benefits affects the development of tourism. For this reason, it is necessary to ensure an equal distribution of income from tourism activities. In this process, it is necessary to give the necessary authority and responsibility to the local communities in the decisions taken about tourism. Local institutions and organizations should be authorized in accordance with the tourism management process.
- In the community-based tourism planning process, incompatibilities between local communities, local businesses, tourism businesses, tourism, and public institutions should be prevented by adopting a participatory planning approach.
- In the regulation of tourism activities, the management of natural resources should be meticulously created, and the harmony of traditional and modern knowledge should be considered in ecological activities.
- Tourism activities should be redesigned to protect endangered populations and species since uncontrolled and unconscious tourism activities pose a serious threat to biological diversity.
- It is necessary to adopt the understanding of sustainability in every step of the tourism movement and to ensure justice between current and future generations.

4.4.3 AGRICULTURAL TOURISM

To benefit from the local people, to be animated and live sensitive, healthy, and organic food, to act in tourism, to talk about life and many types of tourism that will be experienced naturally. One of the tourism types, which has an important place among the new generation tourism activities and supports sustainable development, responsible tourism, and community-based tourism, is agricultural tourism. Agricultural tourism includes economic and/or labor assistance to agricultural and farm works in the regions where tourists travel (Marin, 2015). In the agro–tourism process, participants can participate in daily recreational activities in exchange for accommodation or help with farm work. Participants provide support to production by sharing their agriculture and farming experiences within the scope of the event (Ana, 2017). One of the main purposes of agricultural tourism is to ensure the revival of agriculture and farming and to carry out tourism activities under the umbrella of sustainability in addition to gaining new experiences for the participants. Worldwide Opportunities on Organic Farms (WWOOF) brings together organic farm owners and visitors from around the world. WWOOF is a worldwide movement to create a global community conscious of ecological farming and sustainability practices by providing incentives for visitors to work or financial aid at these farms. WWOOF aims to support sustainable agriculture, develop methods related to organic agriculture and horticulture, and experience rural life by supporting host farmers in its organic farm studies (WWOOF, 2021). Increasing interest in agricultural tourism plays an important role in the regulation of tourism activities with environmental awareness. For this reason, it is necessary for the individuals, participating persons and groups, and farmers who play a role in the regulation of agricultural tourism activities to have received the necessary training in the area (Stanovicic et al., 2018; Iakovidou, 1997).

4.4.4 VOLUNTARY TOURISM

Voluntary tourism activities have been one of the tourism types that gained an important value in the 21st century with the differences in tourism demand. Volunteer tourism has brought a different understanding to tourism by bringing together volunteering activities and travel activities that have experienced rapid development in the tourism sector in the 21st century. Voluntary tourism takes place with the voluntary participation of people in activities for the protection of host communities, destinations and biological life in the host region, in addition to their travels. Volunteer tourism can be carried out simultaneously with all types of special interest tourism. This feature of voluntary tourism enables any tourism consumer to participate in voluntary activities in addition to their activities (Guttentag, 2009). Volunteer tourism travels are made from developed countries to underdeveloped and developing countries around the world. Today, volunteer trips are organized intensively to Asia, Europe, Australia-Pacific, America, Antarctica, and the Middle East. Fields of activity in travel consist of areas such as health, environment, culture, education, and restoration. The objectives of the activity can be listed as meeting the health needs of local communities, providing language and skills training, and raising awareness of individuals for the protection of biodiversity and habitats in the region (Wright, 2013).

4.4.5 ACCESSIBLE TOURISM

Important decisions taken in recent years regarding human rights and freedoms also bear great sensitivity in tourism. A significant part of the world's population cannot participate in tourism activities due to limited opportunities (Bulgan, 2015). Planning all the components that feed tourism at the same level of accessibility for everyone will enable individuals with physical disabilities to take part in tourism activities more easily (Diker and Çetinkaya, 2016). It is thought that it will be possible to provide these opportunities in tourism for disabled individuals, to receive adequate education in society about accessibility and disabled situations, to make a certification system about accessibility for businesses, and to create structures in accordance with universal laws in the neighboring and private service sectors (Akıncı and Sönmez, 2015; Alan, 2013). Again, with technological developments, it has been possible to eliminate disabled people from travel. In the face of this situation, it was deemed necessary to adapt to the latest technological developments in tourism (Jasrotia and Gangotia, 2018). The effort of tourism to adapt to technology has increased with the concept of smart tourism and accessible tourism opportunities (Buhalis and Amaranggana, 2013; Tecau et al., 2019).

4.5 CONCLUSIONS

When the etymology of world tourism is examined, the understanding of sustainability, which has spread especially since the end of the 20th century, has greatly affected the future of tourism. While mass tourism is still effective in the tourism sector, it is known that special interest tourism also creates a serious market. This market, created by a special interest in tourism, has started to attract more attention with the entry of sustainable development into the tourism sector. The aim of gathering special interest tourism types under the roof of sustainable tourism has been effective in redesigning tourism activities with a sustainability understanding. The natural environment, social structure, and historical and cultural assets that are not suitable for the structure of the destinations are seriously damaged, individuals do not have equal conditions in consumption and sufficient physical structure, consumers are in search of new experiences and the number of conscious individuals in consumption increases, touristic enterprises, destination applications and tourism played an important role in reorganizing its activities. In addition to all these, although it is thought that tourism demands consist of mass tourism and special interest tourism, leisure and recreation activities that have gained value in recent years have a great impact on tourism demand. Leisure time is defined as the time that people have left after working, sleeping, and meeting necessary personal chores. Leisure time is also defined as the time when people act according to their wishes. Leisure activities play an effective role in the development of the society in which people live. The COVID-19 global pandemic, which was felt all over the world in 2019, also had serious negative consequences on the tourism sector. The fact that tourism has been an economically important sector since the middle of the 20th century and tourism activities have come to a standstill with the COVID-19 pandemic has once again reminded the importance of recreation and leisure activities (Güneş, 2020). In the research, it has been concluded that together with the changing consumption habits, recreational activities support a healthy life and create social integrity and create tourist attractions with similar effects (Serçek and Serçek, 2015).

KEYWORDS

- **redesigning**
- **reinventing**
- **rebuilding**
- **tourism**

REFERENCES

Abbas, J.; Mubeen, R.; Iorember, P. T.; Raza, S.; Mamirkulova, G. Exploring the Impact of COVID-19 on Tourism: Transformational Potential and Implications for A Sustainable Recovery of the Travel and Leisure Industry. *Curr. Res. Behav. Sci.* **2021,** *2*, 100033.

Adabalı, M. M. Bilinçli Farkındalık Felsefesinin Turizm Alanındaki Yeri: Çatalhöyük Örneği. *Türk Turizm Araştırmaları Dergisi* **2021,** *5* (1), 200–218.

Akıncı, Z.; Sönmez, N. Engelli bireylerin erişilebilir turizm beklentilerinin değerlendirilmesine yönelik nitel bir araştırma. *Anatolia: Turizm Araştırmaları Dergisi* **2015,** *26* (1), 97–113.

Alan, M. Engelli turizmi: Engelliler neden turizm faaliyetlerinde bulunurlar. *Avrupa Sosyal Bilimler Dergisi* **2013,** *39* (3), 480–486.

Ana, M. I. Ecotourism, Agro-Tourism and Rural Tourism in the European Union. *Cactus Tour. J.* **2017,** *15* (2), 6–14.

Antara, M.; Sumarniasih, M. S. Role of Tourism in Economy of Bali and Indonesia. *J. Tour. Hosp. Manage.* **2017,** *5* (2), 34–44.

Arslan, E.; Kendir, H. COVID-19 salgını sonrası yükselen trend kırsal turizm: Zile örneği. *Türk Turizm Araştırmaları Dergisi* **2020,** *4* (4), 3668–3683.

Aydemir, B.; Kazoğlu, İ. H. Toplum temelli turizm: yerel halk algılarını ölçmeye yönelik bir çalışma Halfeti örneği. *Balıkesir Üniversitesi Sosyal Bilimler Enstitüsü Dergisi* **2016,** *19* (35), 169–186.

Aykın, S. M. Avrupa Birliği'nde Sürdürülebilir ve Sorumlu Turizm Uygulamaları ve Türkiye. *Uluslararası Türk Dünyası Turizm Araştırmaları Dergisi* **2018,** *3* (2), 153–170.

Barnes, M. L.; Eagles, P. Examining the Relationship Between Ecotourists and Philanthropic Behaviour. *Tour. Recreat. Res.* **2004,** *29* (3), 35–38.

Blamey, R. K. Principles of Ecotourism. *Encyclopedia Ecotour.* **2001,** 5–22.

Briedenhann, J. Socio-Cultural Criteria for the Evaluation of Rural Tourism Projects–a Delphi Consultation. *Curr. Iss. Tour.* **2009,** *12* (4), 379–396.

Buhalis, D.; Amaranggana, A. Smart Tourism Destinations. In *Information and Communication Technologies in Tourism*; Springer: Cham, 2014; pp 553–564.

Bulgan, G. Dünyada ve türkiye de engelli turizmi ile ilgili yapılan çalışmalar. *Akademik Bakış Uluslararası Hakemli Sosyal Bilimler Dergisi* **2015,** *50*, 102–125.

Butarbutar, R.; Soemarno, S. Environmental Effects of Ecotourism in Indonesia. *J. Indonesian Tour. Dev. Stud.* **2013,** *1* (3), 97–107.

Chang, C. L.; McAleer, M.; Ramos, V. A Charter for Sustainable Tourism After COVID-19. *Sustainability* **2020,** *12* (9), 3671.

Cohen, E. "Alternative tourism"—A Critique. *Tour. Recreat. Res.* **1987,** *12* (2), 13–18.

Çelik, S. Alternatif turizm. *J. Int. Soc. Res.* **2018,** *11* (56).

Diker, O.; Çetinkaya, A. Erişilebilir Turizm Açısından Safranbolu Turizm Destinasyonunun Uygunluğunun Değerlendirilmesi. *Karabük Üniversitesi Sosyal Bilimler Enstitüsü Dergisi* **2016,** *2*, 111–125.

Fang, Y.; Zhu, L.; Jiang, Y.; Wu, B. COVID-19 için ekonomi politikalarının halk sağlığı müdahaleleri kapsamında eğlence ve rekreasyon endüstrisi üzerindeki tahmini etkileri. *Turizmde Güncel Konular* **2021,** 1–13.

Frey, N.; George, R. Responsible Tourism Management: The Missing Link Between Business Owners' Attitudes and Behaviour in the Cape Town Tourism Industry. *Tour. Manag.* **2010,** *31* (5), 621–628.

Giampiccoli, A.; Saayman, M. Community-Based Tourism, Responsible Tourism, and Infrastructure Development and Poverty. *Afr. J. Hosp. Tour. Leis.* **2017,** *6* (2), 1–28.
Guttentag, D. A. The Possible Negative Impacts of Volunteer Tourism. *Int. J. Tour. Res.* **2009,** *11* (6), 537–551.
Güneş, G. Korunan alanların yönetiminde yeni bir yaklaşım: katılımcı yönetim planları. *Ekonomi Bilimleri Dergisi* **2011,** *3* (1), 47–57.
Güneş, S. G. Doğal ve Kültürel Kaynakların Sürdürülebilir Kullanımı Bağlamında Sorumlu Turizm. *Uluslararası Kentleşme ve Çevre Sorunları Sempozyumu* **2018,** *2*, 218–225.
Güneş, S. G. Sorumlu Turizm. In *Sorumlu Turizm* İçinde (s. 133–152); Güneş, S. G., Özdemir, S., Akgül, Eds.; Nobel Akademik Yayıncılık: Ankara, 2020.
Hughes, E.; Scheyvens, R. Corporate Social Responsibility in Tourism Post-2015: A Development First Approach. *Tour. Geogr.* **2016,** *18* (5), 469–482.
Iakovidou, O. Agro-Tourism in Greece: The Case of Women Agro-Tourism Co-Operatives of Ambelakia. *Medit* **1997,** *8* (1), 44–47.
Jaafar, M.; Maideen, S. A. Ecotourism-Related Products and Activities, and the Economic Sustainability of Small and Medium Island Chalets. *Tour. Manage.* **2012,** *33* (3), 683–691.
Jasrotıa, A.; Gangotıa, A. Smart Cities to Smart Tourism Destinations: A Review Paper. *J. Tour. Intell. Smartness* **2018,** *1* (1), 47–56.
Karacaoğlu, S.; Yolal, M.; Bırdır, K. Toplum temelli turizm projelerinde katılım ve paylaşım: Misi köyü örneği. *Çağ Üniversitesi Sosyal Bilimler Dergisi* **2016,** *13* (2), 103–124.
Kozak, M. A.; Evren, S.; Çakır, O. Tarihsel süreç içinde turizm paradigması. *Anatolia: Turizm Araştırmaları Dergisi* **2013,** *24* (1), 7–22.
Köroğlu, Ö.; Karaman, S. Doğaya dayalı turizm faaliyetlerinin gelişiminde toplum temelli doğal kaynak yönetiminin önemi. *Karamanoğlu Mehmetbey Üniversitesi Sosyal Ve Ekonomik Araştırmalar Dergisi* **2014,** *1*, 95–106.
Marin, D. Turizmin ve tarım turizminin yerel topluluklar üzerindeki ekonomik etkisi üzerine çalışma. *Tarım Bilimleri Araştırma Dergisi* **2015,** *47* (4), 160–163.
Mathew, P. V.; Sreejesh, S. Impact of Responsible Tourism on Destination Sustainability and Quality of Life of Community in Tourism Destinations. *J. Hosp. Tour. Manage.* **2017,** *31*, 83–89.
Okazaki, E. A Community-Based Tourism Model: Its Conception and Use. *J. Sustain. Tour.* **2008,** *16* (5), 511–529.
Özçoban, E. Koronavirüs' ün (COVID-19) Turizm Sektörü Üzerindeki Etkileri ve Türkiye'nin Kırsal Turizm Potansiyeli Üzerine Bir Analiz. *Electron. Turk. Stud.* **2020,** *15* (4).
Paskova, M.; Zelenka, J. How Crucial Is the Social Responsibility for Tourism Sustainability? *Soc. Responsibility J.* **2018,** 534–552.
Saarinen, J. Traditions of Sustainability in Tourism Studies. *Ann. Tour. Res.* **2006,** *33* (4), 1121–1140.
Saarinen, J. Sustainable Tourism: Perspectives to Sustainability in Tourism. In *Sustainable Tourism in Southern Africa*, 2009; pp 77–90.
Scoones, I. Sustainability. *Dev. Practice* **2007,** *17* (4–5), 589–596.
Scott, N.; Laws, E.; Prideaux, B. Tourism Crises and Marketing Recovery Strategies. *J. Travel Tour. Market.* **2008,** *23* (2–4), 1–13.
Serçek, S.; Serçek, G. Ö. Turistlerin konaklama işletmesi tercihinde rekreasyon aktivitelerinin etkisi. *Electron. Turk. Stud.*, **2015,** *10* (14), 681–698.
Sheresheva, M. Y. Coronavirus and Tourism. *Population Econ.* **2020,** *4*, 72.

Spenceley, A.; Relly, P.; Keyser, H.; Warmeant, P.; McKenzie, M.; Mataboge, A.; Seif, J. Responsible Tourism Manual for South Africa, Department for Environmental Affairs and Tourism, July 2002. *Responsible Tour. Manual South Africa* **2002,** *2* (3).
Stanovcic, T.; Pekovic, S.; Vukcevic, J.; Perovic, D. Going Entrepreneurial: Agro-Tourism and Rural Development in Northern Montenegro. *Busi. Syst. Res.* **2018,** *9* (1), 107–117.
Sun, D.; Walsh, D. Review of Studies on Environmental Impacts of Recreation and Tourism in Australia. *J. Environ. Manage.* **1998,** *53* (4), 323–338.
Sustainable Tourism. https://www.unwto.org/sustainable-development (Accessed 22 Nov 2021).
Tecau, A. S.; Bratucu, G.; Tescaşiu, B.; Chiţu, I. B.; Constantin, C. P.; Foris, D. Sorumlu Turizm Engelli Çocuklarla Aileleri Turist Destinasyonlarına Entegre Etmek, *Sürdürülebilirlik* **2019,** *11* (16), 4420.
Üner, T. COVID-19 döneminde turizm yazınında rekreasyon çalışmalarının sistematik derleme yöntemi ile incelenmesi. *Int. J. Contemp. Tour. Res.* 5 (Prof. Dr. Özlem İPEKGİL DOĞAN'ı Anma Özel Sayısı) **2021,** 178–189.
Wijaya, I. Sustainable Tourism Concept in Redesigning Zone-Arrangement of Banyuwedang Hot Springs Architecture. *Int. J. Appl. Sci. Tour. Events* **2019,** *3* (1), 54–67.
Wright, H. Volunteer Tourism and It's (Mis) Perceptions: A Comparative Analysis of Tourist/ Host Perceptions. *Tour. Hosp. Res.* **2013,** *13* (4), 239–250.
WWOOF. What Is WWOF? https://wwoof.net/ (Accessed 21 Nov 2021).

CHAPTER 5

Crisis Management and Emergency Response Plan in Tourism During Pandemics and Disasters

ASLI SULTAN EREN[1], ALHAMZAH ALNOOR[2], and GEORGIA TACU[3]

[1] *Intuition of Tourism Research, Tour Guiding, Nevşehir Hacı Bektaş Veli University, Turkey*

[2] *School of Management, MBA, Organizations Studies, Universiti Sains Malaysia, Malaysia*

[3] *Gh. Zane' Institute for Economic and Social Research Iaşi, Romanian Academy, Romania*

ABSTRACT

In this study, information is given about crisis management in tourism enterprise and the importance of implementing an emergency response plan. A crisis can be defined as an event that occurs as a result of an unexpected change in the internal or external environment of the enterprise and threatens the physical and financial condition of the enterprise, its future, and its employees physically and spiritually. Crisis management, on the contrary, can be expressed as determining what needs to be done against the crisis and its effects, making crisis plans, putting them into practice, and evaluating the results. It is known that if the emergency response plan, which does not always have negative effects on businesses, is implemented, it provides important opportunities for tourism businesses and makes them one step ahead of their competitors. The aim of the study is to reveal the importance of the management of crises and emergency response plans, especially due to pandemic diseases.

Dynamics of the Tourism Industry: Post-Pandemic and Post-Disaster Perspectives and Strategies.
Gül Erkol Bayram and Anukrati Sharma (Eds.)

5.1 INTRODUCTION

Societies can fall into chaos for many different reasons. These can be global or regional chaos. These chaotic situations turn into crises over time, resulting in the emergence of a complex structure that becomes more difficult to solve (Aksu, 2009). The crisis is an important event that poses a crucial threat to businesses in every time. Therefore, crisis management should be applied correctly in all types of businesses in order for them to hold on (Bulgan and Aktel, 2017; Demirtas, 2000; Sucu, 2021). Kash and Darling emphasized that success in crisis management is prevention, preparation and correct intervention. It is almost impossible to predict. Therefore, businesses need to be ready for any situation at any time.

In different years, many infectious diseases emerge under the name of epidemics and pandemics. While the disease called epidemic plaque is more effective in and around the region where they occur, the diseases called pandemic are effective worldwide and describe the diseases that cause serious harm to people and cause death (Gül and Çelebi, 2020; Haryanto, 2020; Tunç and Atıcı, 2020). Pandemic diseases, in particular, pose a great threat to different areas. When diseases occur, it may be too late for crisis management. Therefore, businesses should be prepared for these situations (Akyüz, 2020; Güngör, 2020). When a crisis has occurred, businesses should act fast to face least harm (Erkan, 1996). Businesses that have an emergency response plan start crisis management earlier and take the least damage from the crisis process (Cannon et al., 2011). Therefore, the emergency response plan, which is the B plan for businesses, is of great importance. The ability of businesses to hold on in the sector and to overcome the difficult process in the most successful way will help them to be prepared for all crises that may occur in the future (Niska and Burt, 2007). The aim of this study is to moderate the importance of crisis management and emergency response plan implementation in crisis management for businesses. The study will be useful for those who want to have information about the literature and the subject in terms of the information it will give on the fight against the crisis.

5.2 CRISIS MANAGEMENT IN TOURISM DURING PANDEMIC AND DISASTER

Natural disasters, economic problems, terrorism, and political chaos can be counted among the causes of crises that have an impact on many sectors. However, epidemics can also be considered as a cause of the crisis (Çeti and Ünlüönen, 2019). Humanity has struggled with various epidemics, pandemics, and natural disasters throughout history since the time it started to live in communities and continues to do so. During the epidemic, countries tried to develop a number of strategies, tactics, and methods within the scope of combating the epidemic (Tunç and Atıcı, 2020). The ability of businesses to continue their existence depends on how they are protected from these dangers or how they evaluate the opportunities (Asunakutlu et al., 2003).

First, crises are described as a tension-filled process that exceeds the predictive and hindering functions of an organization, threatens the sustainability of strategic goals and operational processes, and is a surprise for management units when it occurs (Özhasar and Ege, 2019; Pira and Sohodol, 2012). A crisis is a situation that prevents a business from holding on to a sector that threatens its goals and requires urgent action; It is a special situation that creates tension when businesses fail to predict the crisis and take precautions (Demirtas, 2000). For an emerging unexpected situation to be described as a "crisis," it must have the following characteristics:

- The crisis situation is unpredictable.
- The organization's forecasting and crisis prevention mechanisms are insufficient.
- The crisis threatens the purpose and existence of the organization.
- Ways to overcome the crisis and follow.
- There is not enough information and time to make a decision.
- The crisis requires immediate action.
- Crisis creates tension among decision makers (Asunakutlu et al., 2003; Seçilmiş and Sarı, 2010).

TABLE 5.1 Factors Causing the Crisis.

Environmental factors	Organizational factors
Economic situation and political system	Change of organizational structure
Technological developments	Quality of management
Change of social and cultural factors	Organization's history and experience
Legal considerations	The life course of the organization
Change of political structure	Inability to collect and use information
Change of international communication	Change of organization culture
Natural factors (disaster)	Coordination problems
Globalization	Communication breakdowns

Source: Adapted from Güngör (2020).

Crises are negativities that cause serious material and moral losses to microscale organizations and macroscale economies of countries. The important thing is to be prepared for the crisis with the awareness that there is always the possibility of the crisis, not only to delay or ward off the crisis but also to find ways to get out of the crisis with the least damage and gain some gains from the crisis (Gül and Çelebi, 2020). Crisis management includes identifying the crisis and its causes, analyzing them, implementing the necessary measures, predicting future crises based on them, implementing the previously prepared plans when crises occur, and taking measures against the crisis (Okumuş, 2003; Seçilmiş and Sarı, 2010). Since crises are an important imbalance situation that destroys the routine activities of the organization, they must be managed with fast, effective, and correct strategies (Akyüz, 2020; Titiz and Çarıkçı, 1990) "Saving the day" is not a crisis management (Sezgin, 2003). Proactive planning and reactive management of crises are one of the critical issues for crisis management (Akyüz, 2020; Bulduklu & Karaçor, 2017). During the pandemic period, important duties fall on both national and international organizations to minimize the negative effects, guide individuals, and manage the emerging crisis correctly (Akyüz, 2020).

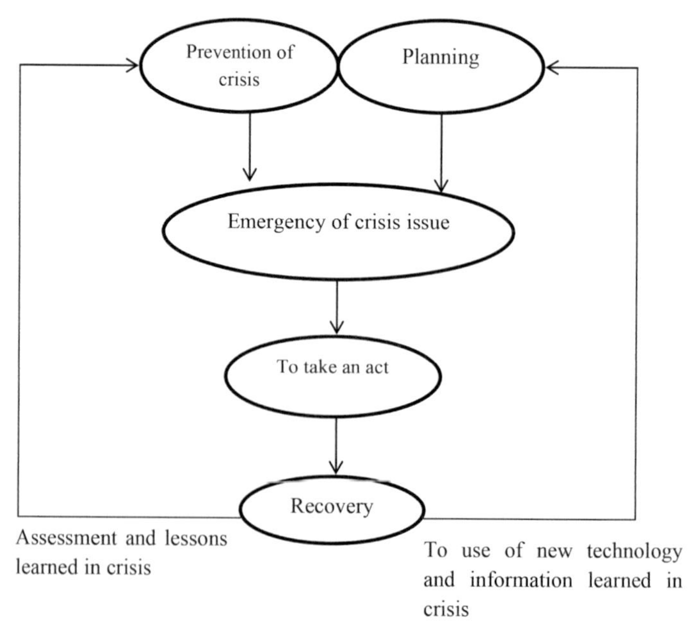

FIGURE 5.1 Crisis management.

Source: Adapted from Brezina and Šimák (2018).

As shown in Figure 5.1, preparedness and crisis planning give the opportunity to intervene immediately in a crisis, and as a result of this reaction, it is seen that even in a crisis, a period of renewal can be reached rapidly. It is necessary to prepare for crises again by using new technologies and to prepare for the next crisis after making the necessary inferences from the crisis process. This process will be an ongoing process for all businesses that successfully manage the crisis (Brezina and Šimák, 2018).

Even the most popular tourist destinations can lose their reputation, attractiveness, and marketability overnight, as the tourism sector is fragile in the face of sudden changes and human or natural events, common diseases, natural disasters, or social events (Beirman, 2003; Çeti and Ünlüönen, 2019).

For the tourism sector, not seeing the signs of crisis is suicide. What is happening in the sector should not be ignored with the logic of "it happened once, it cannot be repeated", and it should immediately seek solutions and should be taken into account. The belief of public administrations that the crisis situation will pass by itself and the thought that the cost of the intervention will be high will adversely affect the resolution of the crisis and the crisis management. If the necessary measures are not taken in time, the losses in tourism revenues will disrupt the balance of payments balance, cause the facilities to be closed or operate with low capacity, and result in an increase in the number of unemployed due to the fact that new facilities are not put into service (Aymankuy, 2001).

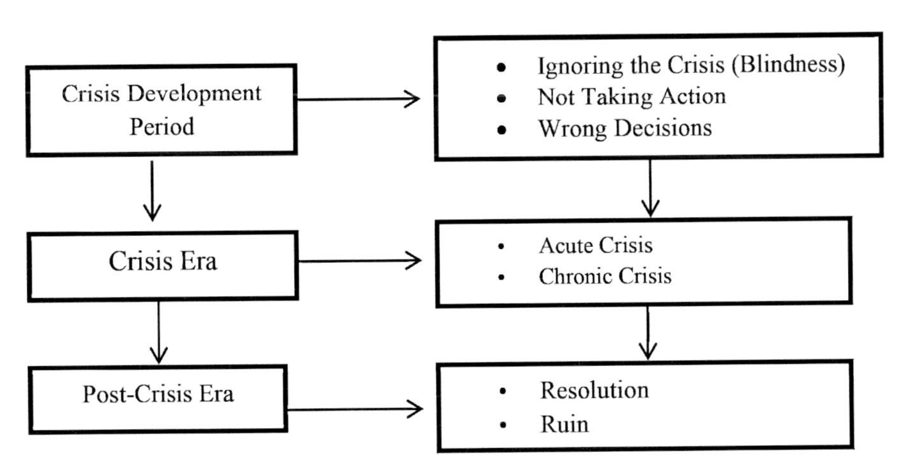

FIGURE 5.2 Crisis management in tourism businesses.

Source: Adapted from Göral (2014).

As shown in Figure 5.2, the development period of the crisis consists of three stages under the headings of blindness, inaction, and making wrong decisions. The first period in which the crisis must be defined is called the blindness stage. At this stage, internal and external changes occur that may threaten the survival of the tourism enterprise, and problems begin to arise in the relations of the enterprise with the environment. The second stage of the development period of the crisis; Despite the decrease in the success rate, there is no activity in the tourism business against this situation, in other words, it is defined as the inactivity phase. In the development period of the crisis, the stage of wrong decision and wrong action refers to the threshold of the crisis. At this last stage, which is on the brink of the crisis, the reason that led the managers to make wrong actions and wrong decisions is accepted as the tendency of individuals with feelings of panic and anxiety to make quick and generally inaccurate decisions instead of creative decisions (Can, H., 2002; Göral, 2014).

The subject of the tourism industry is human, and the crises experienced in the place of people bring along processes that have great effects and are not easy to solve. Since the tourism industry is a sensitive sector against crises, the preparation plans and early warning systems created by the enterprises against crisis management are the most important steps of this process (Çolak and Batman, 2019). It is seen that adaptation to the crisis process and organizational preparation studies come to the fore (Avcı and Küçükusta, 2013).

- It should go into the process of adapting to the new normal.
- The inadequacy of existing functions and areas should be explained.
- New functions and areas should be introduced.

This will bring about a transition period. While the old system is gradually removed, the new system should be installed gradually, that is, the most appropriate parallel transition should be ensured (Topaloğlu and Tunç, 1997). For effective crisis management, it is vital that marketing activities are managed in coordination with human resources, cost, quality, and customer (Yılmaz et al., 2019).

5.3 EMERGENCY RESPONSE PLAN IN TOURISM DURING PANDEMIC AND DISASTER

Emergency response planning is a process involving many stakeholders who can simultaneously communicate through multiple channels and exchange various information structures (Chokshi et al., 2014). Being fast in a crisis situation is vital for the future of businesses. Thanks to the preparation and development of a crisis response plan, the steps to be taken when a crisis arises and the determination (Erkan, 1996) of the duties of each member accelerate the immediate response. A manager needs to understand that crises are inevitable and that failure to manage a crisis effectively and efficiently is due to his negligence, lack of planning, or misunderstanding of the parameters of the crisis. Before a crisis is likely to develop, it is necessary to prepare an emergency response plan and be prepared for a crisis situation (Aksoy and Aksoy, 2003). Supporting preparedness and response activities for crisis management is challenging due to the uncertainty of events and complications resulting from a lack of data. Emergency response plans should be prepared through existing community plans for first response readiness (Campbell et al., 2008).

It will be beneficial for tourism enterprise to establish certain percentages for the steps to be taken according to the importance while preparing the emergency response plan before the onset of the crisis. It will be a faster and more practical solution than looking for a solution immediately in a crisis. It will also be less error-prone and will help ensure a good result rather than a deterministic result which can be misleading (Brown and Dunn, 2007). The importance of developing awareness of the emergency response plan of the managers will help them to achieve successful result as a result of emergency management and to end the crisis period without harming the businesses (Campbell et al., 2008; Tunç and Atıcı, 2020).

Basing emergency planning and response decisions on forecasts and previous crisis experiences can result in significant financial savings and increased safety for both the public and emergency responders (Cannon et al., 2011). The crisis response planning process should be prepared considering that tourism is a sensitive sector, from the decision to prepare it to the implementation stage. In a tourism business with an emergency response plan, crisis management significantly reduces management's blockage and indecision in times of high tension (Özen, 2011). During the pandemics, crisis management will help the entire tourism sector to come out of the pandemic situations that may develop in the future, such as 2020–2021, with the least damage. Emergency response planning will act as a guide for tourism businesses when fighting natural disasters, epidemics, or other crises (Niska and Burt, 2007).

During the occurrence of natural disasters and epidemics, no person or institution is on the scene. Only the frightening hum of natural disasters is heard on the stage, the destruction is seen, or the people or countries that lost the war with epidemics are in the middle (Aydın and Güner, 2013). When the disaster stops, the curtain is closed, but the aftereffects and durations of the aftereffects are longer. This increases the human, economic, cultural, and natural losses in tourism, which is still seen as a "luxury" for people, but provides job opportunities for thousands and millions of people. The first actors to appear on the stage right after the end/stop of the risk are the actors of the "Emergency Response Practice" planned before the risk. In the risk management phase, which is implemented before the risk, high-level training should be given to the employees working in tourism enterprise in a way that the damage foreseen in the planning will not be exceeded and that they can calculate the losses in the planning.

5.4 CONCLUSIONS

Crises are the integral part of our life. Epidemics, natural disasters, wars, terrorist incidents, or socioeconomic crises have occurred in every period of history. Such crises will continue in the future. As a matter of fact, since the end of 2019, the world has been struggling with COVID-19. In other words, it is possible to say that we are currently in a crisis situation.

The number of people traveling internationally across the world is approximately 375 million by the end of 2021, which is 75% less compared to 2019 (about 1.4 billion). Compared to Asia and the Pacific (−95%), Africa (−74%), the Middle East (−81%), Europe (−53%) and the USA (−60%) suffered less in terms of the decline in tourist numbers. Although tourism revenues will increase slightly compared to 2020 and reach 707 billion euros in 2021, it is expected to be less than half of 1.5 trillion euros in 2019 (Bolelli, 2021). The figures reveal the seriousness of the crisis. After the COVID-19 epidemic, many accommodation businesses, restaurants, and businesses serving different areas of tourism were closed. Crisis management is an important practice that enables us to continue on the road by taking the least possible damage from such disasters.

The most important practices are the adaptation of business processes to new conditions and the adoption of a flexible and rapidly adapting management approach, in which new approaches should be adopted by reviewing the management. In addition, one of the most important negative consequences of the crisis is the increase in the level of stress and difficulties in payments and receivables. Management's realistic approach and organizational methods of coping with stress are among the practices in crisis environments where the entire sector is affected. In the crisis, travel businesses should also consider the practices of competing with quality, creating strategic partnerships, conducting customer loyalty studies, creating new sources of income, and entering new markets.

The emergency response plan is another important application. The emergency response plan can be an important competitive opportunity for tourism businesses in unexpected but long-lasting crises such as illness. It is possible to say that the hotels, which developed an emergency response plan against epidemics, applied the distance and hygiene rules before the governments took decisions such as restrictions, and thus they passed through this crisis period with the least possible damage.

The period we have been in for the last 2 years shows that the crisis is inevitable and must be fought. Although the crisis often has negative consequences, those who are prepared always manage to turn the crisis into an opportunity. Travel agencies with an emergency response plan continued to accept tourists by halving their capacity, while those without an emergency response plan closed. Similar to this, we can observe how many tourism stakeholders got ahead of their competitors and went through the crisis period.

Being prepared for any situation is the key to success. It is possible to successfully overcome the next crisis period with the lessons learned from a crisis that brought bad results. A crisis is an inevitable reality that existed in every aspect of human life yesterday, exists today, and will exist tomorrow. Ignoring is not the right answer. Fear is the worst. The only way to successfully exit from the crisis is to struggle.

KEYWORDS

- **crisis management**
- **emergency response plan**
- **tourism**
- **pandemic disaster**

REFERENCES

Aksoy, H. H.; Aksoy, N. OkullardaKrize Müdahale Planlamasi. *Ankara Univ. J. Faculty Educ. Sci.* **2003,** *36* (1–2), 37–49. https://doi.org/10.1501/egifak_0000000076

Aksu, A. Kriz Yönetimi ve Vizyone Liderlik. *J. Yasar Univ.* **2009,** *4* (15), 2435–2450. https://doi.org/10.19168/jyu.45318

Akyüz, M. Crisis Management Arrroaches of Health Service OOrganizations in The Context of The COVID-19 Pandemic: Comparisons and a Model Proposal. *J. Strategic Manage. Res.* **2020,** *3* (2), 159–178. http://www.nutricion.org/publicaciones/pdf/prejuicios_y_verdades_sobre_grasas.pdf%0Ahttps://www.colesterolfamiliar.org/formacion/guia.pdf%0Ahttps://www.colesterolfamiliar.org/wp-content/uploads/2015/05/guia.pdf

Asunakutlu, T.; Safran, B.; Tosun, E. Kriz Yönetimi Üzerine Bir Araştırma. *Dokuz Eylül Üniversitesi Sosyal Bilimler Enstitüsü Dergisi* **2003,** *5* (1), 141–163.

Avcı, N.; Küçükusta, D. Effect of Global Economic Crisis on Travel Agencies And Crisis Management Practices in Turkey. *Dokuz Eylül Üniversitesi Sosyal Bilimler Enstitüsü Dergisi* **2013,** *15* (4), 571–587.

Aydın, M.; Güner, S. Risk Management in Cultural Heritage. *Batman Univ. J. Life Sci. Vol.* **2013,** *3* (1), 70–81. https://dergipark.org.tr/tr/pub/buyasambid/issue/29821/320787

Beirman, D. *Restoring Tourism Destinations in Crisis*; Allen & Unwin, 2003.

Bolelli, Ş. *COVID-19 Continued to Hit the Tourism Industry all Over the World in 2021*; Anatolian Agency, 2021.

Brezina, D.; Šimák, L. Proposal for a Model of Effective Reaction To Natural Disasters in the Territory of Slovakia. *CBU International Conference Proceedings* **2018,** *6* (February), 1017–1023. https://doi.org/10.12955/cbup.v6.1288

Brown, D. F.; Dunn, W. E. Application of a Quantitative Risk Assessment Method to Emergency Response Planning. *Comput. Operat. Res.* **2007,** *34* (5), 1243–1265. https://doi.org/10.1016/j.cor.2005.06.001

Bulduklu, Y.; Karaçor, S. Crisis Communication and New Media in Health Services. *Atatürk İletişim Dergisi* **2017,** *14*, 279–296. https://dergipark.org.tr/en/pub/ataunniletisim/issue/34005/357162

Bulgan, G.; Aktel, M. A Resaerch on Crisis Management of Five Star Hotels in Antalya. *Dokuz Eylül Üniversitesi İşletme Fakültesi Dergisi* **2017,** *18* (2), 205–232. https://doi.org/10.24889/ifede.306868

Campbell, B. D.; Mete, H. O.; Furness, T.; Weghorst, S.; Zabinsky, Z. Emergency Response Planning and Training Through Interactive Simulation and Visualization with Decision Support. *2008 IEEE International Conference on Technologies for Homeland Security, HST'08*, 2008; pp 176–180. https://doi.org/10.1109/THS.2008.4534445

Can, H. *Organizasyon ve Yönetim*; Siyasal Kitabevi, 2002.

Cannon, S. H.; Boldt, E. M.; Laber, J. L.; Kean, J. W.; Staley, D. M. Rainfall Intensity-Duration Thresholds for Postfire Debris-Flow Emergency-Response Planning. *Natural Hazards* **2011,** *59* (1), 209–236. https://doi.org/10.1007/s11069-011-9747-2

Çeti, B.; Ünlüönen, K. Evaluation of the Effect of Crisis Due to Epidemic Diseases on the Tourism Sector. *AHBVU J. Tour. Faculty* **2019,** *22* (2), 109–128. https://doi.org/10.34189/tfd.22.2.001

Chokshi, A.; Seyed, T.; Rodrigues, F. M.; Maurer, F. EPlan multi-surface: A multi-surface Environment for Emergency Response Planning Exercises. *ITS 2014—Proceedings of the 2014 ACM International Conference on Interactive Tabletops and Surfaces* 2014; pp 219–228. https://doi.org/10.1145/2669485.2669520

Çolak, O.; Batman, O. Crisis Management in Tourism: The Case of İstanbul. *Saffron J. Cult. Tour. Res.* **2019,** *2* (3), 351–371. https://dergipark.org.tr/tr/pub/saktad/574620

Demirtas, H. Kriz Yönetimi Makale. *Kuram ve Uygulamada Eğitim Yönetimi* **2000,** *23* (23), 353–373.

Erkan, S. Krize müdahale planı. *Eğitim Yönetimi* **1996,** *2* (1), 547–554.

Göral, R. A Holistic Framework Related To Strategic Crisis Management In Tourism Sector. *Selcuk Univ. J. Inst. Soc. Sci.* **2014,** *32*, 89–101. https://search.proquest.com/docview/1625941931?accountid=28391

Gül, H.; Çelebi, F. Koronavirüs (COVID-19) Pandemisinde Başlıca Gelişmiş ve Gelişmekte Olan Ülkelerde Kriz Yönetiminin Değerlendirilmesi. *MANAS J. Soc. Students* **2020,** *9* (3), 1703–1715. https://doi.org/10.33206/mjss.719164

Güngör, B. Evaluation of Measures Taken During the COVID-19 Pandemic From Crisis Management Perspective in Turkey. *Int. J. Soc. Sci. Acad.* **2020,** *2* (4), 818–851.

Haryanto, T. Editorial: COVID-19 Pandemic and International Tourism Demand. *Journal of Developing Economies* **2020,** *5* (1), 1. https://doi.org/10.20473/jde.v5i1.19767

Kash, T. J.; Darling, J. R. Crisis Management: Prevention,Diagnosis and Intervention. *Leadership Organiz. Dev. J.* **1998,** *19* (4), 179–186.

Niska, R. W.; Burt, C. W. Emergency Response Planning in Hospitals, United States: 2003–2004. *Adv. Data* **2007,** *391*, 1–13.

Okumuş, F. İşletmelerde Kriz Yönetimi ve Krizlerin İşletmeler Üzerine Olası Etkileri. *Atatürk Üniversitesi İktisadi ve İdari Bilimler Dergisi* **2003,** *17* (1–2), 203–212. https://doi.org/10.16951/iibd.06113

Özen, Y. Managerial Planning and Training Educators and Individual Crisis Itervention. *Gümüşhane Üniversitesi Sosyal Bilimler Enstitüsü Dergisi* **2011,** *2* (3), 119–140.

Özhasar, Y.; EGE, Z. Crisis Management Practices In Accommodation Enterprises: Sample Of Istanbul Fatih District. *Int. J. Contemp. Tour. Res.* **2019,** *3*, 50–65. https://doi.org/10.30625/ijctr.545435

Pira, A.; Sohodol, Ç. *Kriz Yönetimi Halkla İlişkiler Açısından Bir Değerlendirme*, 4th ed; İletişim Yayınları, 2012.

Seçilmiş, C.; Sarı, Y. A Study on The Crisis Management Practiceses of Acoommodation Establishment DUuring Crisis Periods. *Suleyman Demirel Univ. J. Faculty Econ. Admin. Sci.* **2010,** *15* (1), 501–520.

Sezgin, F. Kriz yönetimi. *Kırgızistan Türkiye Manas Üniversitesi Sosyal Bilimler Dergisi* **2003,** *8* (1), 181–195. https://doi.org/10.14527/9786053640677.17

Sucu, M. Crisis Management in the Aviation Sector During the COVID-19 Pandemic: A Case Study for Turkish Airlines (THY). *J. Busi. Res. – Turk* **2021,** *13* (3), 2863–2884. https://doi.org/10.20491/isarder.2021.1295

Titiz, İ.; Çarıkçı, H. İ. Krizlerin İşletmeler Üzerindeki Etkileri ve Küçük İşletme Yöneticilerinin Kriz Dönemine Yönelik Strateji Analizleri. *C.Ü. İktisadi ve İdari Bilimler Dergisi* **1990,** *2* (1), 203–218.

Topaloğlu, M.; Tunç, A. Turizm İşletmelerinde Kriz Yönetimi. *Anatolia: Turizm Araştırmaları Dergisi* **1997,** *8* (1), 88–94.

Tunç, A.; Atıcı, F. Z. Struggling Pandemics in the World and in Turkey : A Study in the Context of Risk and Crisis Management. *TROYACADEMY Int. J. Soc. Sci.* **2020,** *5* (2), 329–362.

Yakut Aymankuy, Ş. Crisis Management in Tourism. *Balıkesir Üniversitesi Sosyal Bilimler Dergisi* **2001,** *6*, 11.
Yılmaz, Y.; Ünal, C.; Dursun, A. Crisis Management Practices and Predictions of Hotel Managers Concerning the 2016 Tourism Crisis in Turkey. *J. Yasar Univ.* **2019,** *14* (56), 468–488.

CHAPTER 6

The Recovery Roadmap of the Tourism Industry Post-COVID-19 Pandemic: Literature-Based Recommendations

FARHAD NAZIR[1], JEETESH KUMAR[2], NEETHIAHNANTHAN ARI RAGAVAN[3], and ANUKRATI SHARMA[4]

[1]*University of Coimbra, CEGOT, Faculty of Arts & Humanities, Portugal; University of Swat, Pakistan*

[2]*School of Hospitality, Tourism and Events; Centre for Research and Innovation in Tourism (CRiT); Sustainable Tourism Impact Lab, Taylor's University, Malaysia*

[3]*Faculty of Social Sciences & Leisure Management, Taylor's University, Malaysia*

[4]*Department of Commerce and Management, University of Kota, Kota, India*

ABSTRACT

In a very inclusive global business environment, a strong inclination toward the financial and employment dependence on the tourism industry exists quite visibly. Statistics revealed that every tenth job had been directly or indirectly associated with tourism and its ancillary segments. It is safe to argue that the economies of the specific countries entail a large chunk of revenue generated from the transactions of the tourism business. Considering this, a competitive setup has been created to have a more significant input to national economies due to tourism. Purposeful infrastructural development has been carried out in certain countries to dominate the tourism industry. Such maximum reliance

Dynamics of the Tourism Industry: Post-Pandemic and Post-Disaster Perspectives and Strategies.
Gül Erkol Bayram and Anukrati Sharma (Eds.)

has undergone unprecedented negative impacts, particularly in the ongoing emergency of the COVID-19 pandemic. Pandemic has introduced exemplary restrictions and stoppages to the tourism sector, completely halting the entire hustling happenings to the freezing zone. Unmatchable decline due to pandemics has caused massive economical outbursts and the loss of countless jobs. Furthermore, uncertainty regarding the mitigation and recovery toward the new normal exists due to the frequent prevailing layers of pandemic and novel mutated variants. However, this is not the first case of its kind, and the past literature shows that such health pandemics and disasters have occurred over time. Besides the devastating impacts of these pandemics and disasters, recovery of the tourism industry has been successfully planned and achieved. Embarking on the same, the current study highlights the recovery strategy of the tourism industry in the face of the COVID-19 pandemic through the extensive review of past studies. The archival analysis of the existing studies has enabled the researcher to set implications and recommendations for the speedy recovery of the tourism industry.

6.1 INTRODUCTION

The lucrative face of tourism has attracted the attention of investors and shareholders, eventually transforming this sector into a mega-industry (Camilleri, 2018). In addition to several other ancillary subsectors, hospitality and travel segments have become prominent, generating enough revenue (Thommandru et al., 2021; Vogel, 2021). A large chunk of revenue to the domestic, national, and regional economies have obligated the stakeholders to pool up and claim their shares from the transactions of tourism businesses. However, this hustling industry has a flip side in the form of natural calamities, natural health emergencies, and human-induced unrest and uprising situations (Cohen, 2012; Ozbay et al., 2021; Perles-Ribes et al., 2018; Ritchie, 2004). These unwanted circumstances have impacted tourism's happenings over a long period in the past and eventually left financial, social, cultural, environmental, and psychological impacts. Even the post-events scenarios have created negative imagery of destinations, eventually posing a solid barrier for potential tourists (Avraham, 2006; Avraham and Ketter, 2017; Li et al., 2018; Mair et al., 2016; Ryu et al., 2013; Selmi and Dornier, 2017; Wang, 2017; Wu and Shimizu, 2020). Scaling from least developed–developing–developed and up to the postdeveloped countries has further differences in tackling the impacts of such natural and human-induced unrest situations. Developed and postdeveloped countries seem to be more vigilant and swift in assessing and analyzing the unrests situations (Blake and Sinclair, 2003; Chan et al., 2020; de Sausmarez, 2007; Pennington-Gray et al., 2011; Ritchie, 2008; Ritchie et al., 2004; Zeng et al., 2005). Moreover, these countries' mitigation and recovery strategy has been designed on a war footing, allowing them to recover from such situations and roll back the previous stages of tourism happenings. While in developing and underdeveloped countries, the two-fold consequence patterns of these situations seem to be a big challenge. In the first instance, such countries have to face the undesirable negative consequences of these situations. Later, in the post-events scenarios, a significant challenge of a negative image in the global face regarding threats and risks awaits the stakeholders. Overall to say, despite such devastating impacts, destination's stakeholders have struggled hard to regain their desired image through rebranding and other marketing mantras. To this end, utmost attention to mitigating these undesired situations and recovery strategy has been nominated as a powerful compulsion. Successful mitigation campaigns and recovery roadmaps have been structured in many countries, easing the tourist's decision-making process (Glyptou, 2021; Mansfeld, 1999; Pizam, 2006; Santana, 2004).

Interestingly, in the ongoing globalised milieu, even the developed countries count on mitigating impacts and recovery from pandemics in developing and least developed countries. Such dependence is the cross-overs of nations and exchange of industrial products and services among all scales of development. The unending pandemic—COVID-19 and its fast spread has justified this dependency and urged the shareholders to think as a whole rather than on a solitary and country-specific approach. Keeping with the current discussion, this study has attempted to unveil the theoretical endeavors and organizational frameworks crafted during the pandemic; also, comparatively with the past examples of inevitable pandemics and health emergencies which had affected the tourism industry. Doing so, in both academic and industrial discourse, through this study, an attempt has been made to sketch the potential roadmap for the recovery of tourism from such health outbreaks.

6.2 TOURISM INDUSTRY GLOBALLY

Tourism is a significant source of foreign exchange earnings, a generator of personal and corporate incomes, a creator of employment, and a contributor to government earnings. It is a dominant international activity, surpassing even trade oil and made products. International tourist arrivals grew 5% in

2018 to reach the 1.4 billion mark. This figure was reached 2 years ahead of the UNWTO forecast. At the same time, export earnings generated by tourism have grown to USD 1.7 trillion (UNWTO, 2021). This makes the sector a truly global force for economic growth and development, driving more and better jobs and serving as a catalyst for innovation and entrepreneurship. In short, tourism is helping build better lives for millions of individuals and transforming whole communities. Growth in international tourist arrivals and receipts continues to outpace the world economy, and emerging, and advanced economies benefit from rising tourism income. For the seventh year in a row, tourism exports grew faster than merchandise exports, reducing trade deficits in many countries.

Tourism is an essential component of export diversification for emerging and advanced economies, with a solid capacity to reduce trade deficits and compensate for weaker export revenues from other goods and services. There was an extra USD 121 billion in export revenues from international tourism (travel and passenger transport) for 2018. Export earnings from international tourism are an essential source of foreign revenues for many destinations in the world (UNWTO, 2019, 2020).

Having the 10th consecutive year of sustained growth, international tourist arrivals in 2019 touched 1.5 billion and millions of jobs with a high

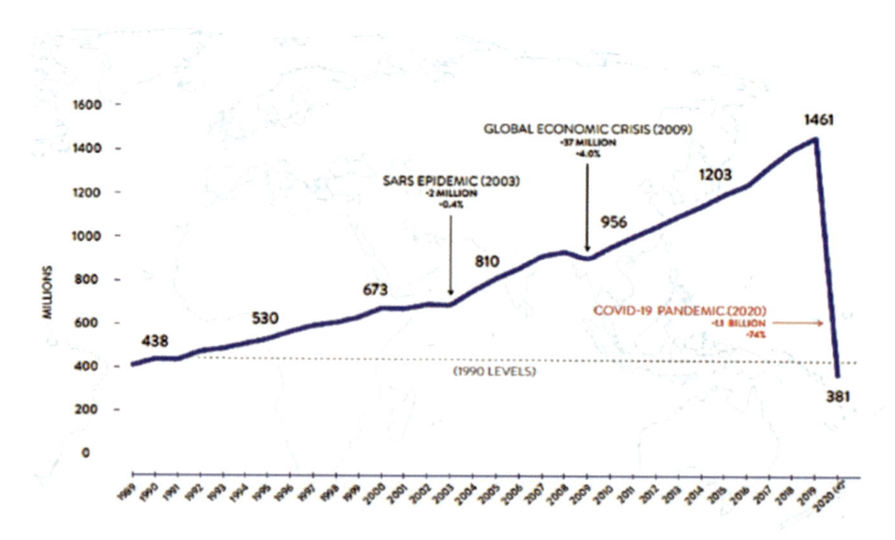

FIGURE 6.1 International tourists arrival (millions).

Source: Reprinted from UNWTO (2021).

share of women. In 2020, a vast downfall (−74%) was recorded for tourists' arrival because of COVID-19 forwarding back international tourism to the level of 30 years ago. This loss is too huge, including 1 billion international tourist arrivals, US$ 1.3 trillion export revenue, over US$ 2 trillion global GDP and 100–120 million direct tourism jobs are also at risk. The UNWTO world tourism barometer report confirms that most tourism experts expect international tourism to return to pre-COVID levels only in 2023 or after (UNWTO, 2021).

6.3 COVID-19 AND TOURISM INDUSTRY

It is safe to state that the pandemic has affected the tourism industry almost in every geographical position. Every development in the face of the pandemic has been directly considered associated with the tourism industry. Vaccinated passports, automated security checks, and contactless hospitality services have all seen significant development and market demand due to changes in the behavior of tourists. National and international Destination Management Organizations DMOs have been entirely engaged in understanding these new tourists and a competitive environment created among several tourism subsectors to market the destinations as safer and more hygienic. Of course, in the post-pandemic era, the preferences of tourists and responsive products and services will have to be entirely dependent on these new health mantras. To furnish such health protocols and standards better, it is necessary to understand the nature of the pandemic's impacts on the tourism industry.

6.4 EFFECTS OF COVID-19 ON THE TOURISM INDUSTRY

Layers of COVID-19 have brought some of the worst and unmatchable consequences ever imagined and least expected, primarily for the supply and demand side of the tourism industry. The impact of a pandemic may be understandable in four major segments of the society, namely, financial, social, cultural, and psychological. Studies based on different geographical domains and stages of development have corroborated the impacts of a pandemic on these segments (Adom, 2020; Bukuluki et al., 2020; Craig, 2020; Renzaho, 2020). The financial crises chain has witnessed some of the most devastating impacts ranging from households to the national, regional, and global economic levels. The trade has suffered financial setbacks due to the pandemic (Mtimet et al., 2021). Severe damage has handicapped the countries, particularly the developing ones disabling them to support the tourism sector. Besides, the developed countries have fueled a considerable number of financial resources to sustain several tourism sectors. For developing and least developed ones, this pandemic has compelled the states to shift on the backfoot and emphasize the essential segments (food, health, shelter) of the community, neglecting the importance of tourism and its ancillary sectors.

While arguing on the grounds of social and cultural frameworks, the pandemic has introduced some unexpected consequences and measures, disintegrating the populace ranging from tiny community segments to the significant global stage. Tourism is the reason for sociocultural get-togethers, maybe well-argued as the frontier sector in these effects. Stoppages of traveling and cancelation of tourism activities compromised sociocultural experiences and the financial shortages (Cooper and Alderman, 2020). Primarily to say, the pandemic has halted the phases of cultural exchange, including the cultural exposure and cultural awareness between the guest and host communities (Swaikoski, 2020: 2). Psychological wellbeing, an additional element of the human health system, has also suffered dramatically during the pandemic (Chua et al., 2021). The psychosomatic problems have surfaced either on the demand side or supply due to the pandemic's immense financial and social pressure. On side of demand avenue, pandemic has entirely transformed the psychological needs of tourists (Cheung et al., 2021), while on the supply side, employees' psychological capital suffered on a larger scale (Mao et al., 2020).

6.5 COVID-19: AN OPPORTUNITY FOR INNOVATION OR A CRISES?

Undoubtedly, the pandemic has been and is expected to be a crisis in an exceptionally pretty long time ahead. However, it has also provided the shareholders and stakeholders of the tourism industry a pause and think time to reconsider their policies, operational setups, and safety procedures. During tourism, prior to pandemic, it was hard to sit back and reassess and reconsider the operational strategies in the past. Voluminous of academic endeavors have corroborated the association of pandemic and tourism more critically and highlighted the potential strategies for the endurance of the tourism industry (Seyfi et al., 2020; Yeh, 2021). Imperative to state the initiation of virtual gadgets while following the space restrictions and contactless compulsions have entirely refurbished the tourism industry.

6.6 CRISIS AND DISASTER MANAGEMENT

Considering this unhappy bond (tourism and crises/disasters) national and international agencies have been mobilized to counter and manage such unwanted situations. In academic discourse, the new terminology of tourism crises and disaster management (TDCM) has been introduced to explicitly address this sensitising element (Jiang et al., 2019). Based on the past literature, TCM (Theory-Context-Method) has proposed potential areas to be emphasized in the face of crises and disasters: theories of crises prevention and preparedness, risk communication, crises management education and training, risk assessment, crises events in the context of COVID-19, data privacy in hospitality and tourism, political-related crises events, digital media, and lastly the analytical methods and approaches (Wut et al., 2021). Similarly, learning from the past unwanted happenings and devising the coping and recovery strategies in a segregated framework has been documented as another viable approach in the face of crises and disasters in the tourism industry (Faulkner, 2001). From pre-crisis planning to implementation and feedback, the crises and disaster management in the tourism business has been argued as another proactive methodology (Ritchie, 2004). Combining the tourism business's sectors, an integrated line of action has been commented as adoptive and result-oriented in enabling the destination to recover from crises impacts (Hystad and Keller, 2008). Primarily, in the milieu of the pandemic, the engagement of multiple players and benefiting from their dynamic capabilities have been warranted as a deemed necessary approach for the sustenance of tourism in crises and disasters (Jiang et al., 2021).

6.7 INTERDISCIPLINARY APPROACHES

Considering the face of pandemic and involvement of different sectors to combat and curb its spread, it is safer to claim that pandemic has brought all these actors on the same schedule. In tourism, the pandemic has introduced some of the novel dependency of health and travel sectors. The mobility of domestic and cross-bordered tourists has been consistently relying on the national and international health agencies. This cross-disciplinary approach has pooled up tourism and health stakeholders, creating a new slogan of collective wisdom to tackle the pandemic (Wen et al., 2021). Health precautions including quarantine, COVID-Test, and vaccination have now taken a position in almost every state's travel prerequisite. Assuredly, this pandemic has launched a paradigm shift in the tourism industry through a synergy of the health sector, media, medical academia, and many others.

6.8 CONCLUSIONS

In a more uncertain global scene, unwanted natural or human-induced happenings have impacted tourism and ancillary segments more than any other sector. Pandemic is one of the most suitable to quote in this regard, deserting the great hustle of tourism business to almost total stoppage. From local to national and regional to global, the financial loss has been immense and expected to continue before the new normal may introduce specific silver linings. The COVID-19 pandemic has wreaked havoc on the tourism industry, causing significant employment and business losses. The tourist industry was one of the first to be struck by the pandemic since measures to contain the virus resulted in a near-complete stoppage of tourism operations worldwide. With the current travel restrictions and the gloomy outlook, the sector risks becoming one of the last to recover. Each of the restrictions to curb the pandemic has been a significant setback for the functionality of tourism activities, despite the availability of management plans and policies at some destinations. The unprecedented launch of social distancing, travel closures, and standard operating procedures (SOPs) has caused the least anticipated consequences ever. Despite all these and past experiences of eventualities, the tourism sector seeks to recover from this global health crisis and longs for the new normal. This study has illuminated the theoretical contexts crafted for the management and recovery of crises and disasters in the tourism industry. Extensively, the existing literature has been brainstormed and discussed to unveil the pragmatic dimensions proposed in academia.

The government should improve its long-term recovery plan by addressing the lack of social protection in the tourism sector to mitigate the impact of a predicted downsizing. The government will be better positioned to deal with private actors after the social safety net is constructed. As we accept the new normal of increased technology disruption and drastic shifts in consumer behavior, the government should address the lack of innovation inside the industry. As a result, the government would have lowered industry players' fear of competition. They will resume making new investments as the tourist industry seizes chances created by changes in travel patterns and trends.

The government should think about adopting policies that provide more extensive social protection for tourism industry participants. In the tourism industry, the percentage of unorganized and seasonal labor is substantially higher. Within the new average, social protection should assist businesses and workers in adapting to new markets. By providing business loans and reskilling programmes, they can assist these players in adopting different business strategies and employment. As a result, strong welfare measures could be implemented to protect their livelihoods in times of uncertainty.

The government could consider fostering an environment in which businesses compete on an equal footing. For example, all industry players should be treated similarly regarding taxation policy and municipal regulations. On the other hand, any rules should be balanced, considering firm growth, and creating a level playing field. The government's involvement in the tourism industry should also be reconsidered. Instead of immediately competing with enterprises, it can assist Small Medium Enterprises SMEs in redeveloping their businesses to meet current and future trends. The government can transfer money to SMEs for programmes such as workforce upskilling.

KEYWORDS

- **tourism industry**
- **pandemic**
- **disasters**
- **mitigation**
- **recovery**
- **restrictions**
- **COVID-19**

REFERENCES

Adom, D. The COVID-19 Global Pandemic: Socio-Cultural, Economic and Educational Implications from Ghana. *Int. Multidisc. J. Soc. Sci.* **2020,** *9* (3), 202–229. https://doi.org/10.17583/rimcis.2020.5416

Avraham, E. Public Relations and Advertising Strategies for Managing Tourist Destination İmage Crises. In *Tourism, Security and Safety*; Routledge, 2006; pp 240–256.

Avraham, E.; Ketter, E. Destination İmage Repair While Combatting Crises: Tourism Marketing in Africa. *Tour. Geogr.* **2017,** *19* (5), 780–800. https://doi.org/10.1080/14616688.2017.1357140

Blake, A.; Sinclair, M. T. Tourısm Crısıs Management: Us Response to September 11. *Ann. Tour. Res.* **2003,** *30* (4), 813–832. https://doi.org/10.1016/S0160-7383 (03)00056-2

Bukuluki, P.; Mwenyango, H.; Katongole, S. P.; Sidhva, D.; Palattiyil, G. The Socio-Economic and Psychosocial İmpact of COVID-19 Pandemic on Urban Refugees in Uganda. *Soc. Sci. Human. Open* **2020,** *2* (1), 100045. https://doi.org/10.1016/j.ssaho.2020.100045

Camilleri, M. A. The Tourism Industry: An Overview. In *Travel Marketing, Tourism Economics and the Airline Product: An Introduction to Theory and Practice*; Camilleri, M. A., Ed.; Springer International Publishing, 2018; pp 3–27. https://doi.org/10.1007/978-3-319-49849-2_1

Chan, C.-S.; Nozu, K.; Cheung, T. O. L. Tourism and Natural Disaster Management Process: Perception of Tourism Stakeholders in the Case of Kumamoto Earthquake in Japan. *Curr. Iss. Tour.* **2020,** *23* (15), 1864–1885. https://doi.org/10.1080/13683500.2019.1666809

Cheung, C.; Takashima, M.; Choi, H.; Yang, H.; Tung, V. The impact of COVID-19 Pandemic on the Psychological Needs of Tourists: Implications for the Travel and Tourism İndustry. *J. Travel Tour. Market.* **2021,** *38* (2), 155–166. https://doi.org/10.1080/10548408.2021.1887055

Chua, B.-L.; Al-Ansi, A.; Lee, M. J.; Han, H. Impact of Health Risk Perception on Avoidance of İnternational Travel in the Wake of a Pandemic. *Curr. Iss. Tour.* **2021,** *24* (7), 985–1002. https://doi.org/10.1080/13683500.2020.1829570

Cohen, E. Globalisation, Global Crises and Tourism. *Tour. Recreat. Res.* **2012,** *37* (2), 103–111. https://doi.org/10.1080/02508281.2012.11081695

Cooper, J. A.; Alderman, D. H. Cancelling March Madness Exposes Opportunities for a More Sustainable Sports Tourism Economy. *Tour. Geogr.* **2020,** *22* (3), 525–535. https://doi.org/10.1080/14616688.2020.1759135

Craig, D. Pandemic and İts Metaphors: Sontag Revisited in the COVID-19 Era. *Eur. J. Cult. Stud.* **2020,** *23* (6), 1025–1032. https://doi.org/10.1177/1367549420938403

de Sausmarez, N. The Potential for Tourism in Post-Crisis Recovery: Lessons from Malaysia's Experience of the Asian Financial Crisis. *Asia Pac. Busi. Rev.* **2007,** *13* (2), 277–299. https://doi.org/10.1080/13602380601045587

Faulkner, B. Towards a Framework for Tourism Disaster Management. *Tour. Manage.* **2001,** *22* (2), 135–147. https://doi.org/10.1016/S0261-5177 (00)00048-0

Glyptou, K. Destination Image Co-creation in Times of Sustained Crisis. *Tour. Plan. Dev.* **2021,** *18* (2), 166–188. https://doi.org/10.1080/21568316.2020.1789726

Hystad, P. W.; Keller, P. C. Towards a Destination Tourism Disaster Management Framework: Long-Term Lessons from a Forest Fire Disaster. *Tour. Manage.* **2008,** *29* (1), 151–162. https://doi.org/10.1016/j.tourman.2007.02.017

Jiang, Y.; Ritchie, B. W.; Benckendorff, P. Bibliometric Visualisation: An Application in Tourism Crisis and Disaster Management Research. *Curr. Iss. Tour.* **2019,** *22* (16), 1925–1957. https://doi.org/10.1080/13683500.2017.1408574

Jiang, Y.; Ritchie, B. W.; Verreynne, M.-L. Building Dynamic Capabilities in Tourism Organisations for Disaster Management: Enablers and Barriers. *J. Sustain. Tour.* **2021,** 1–26. https://doi.org/10.1080/09669582.2021.1900204

Li, F.; Wen, J.; Ying, T. The İnfluence of Crisis on Tourists' Perceived Destination İmage and Revisit İntention: An Exploratory Study of Chinese Tourists to North Korea. *J. Destination Market. Manage.* **2018,** *9*, 104–111. https://doi.org/10.1016/j.jdmm.2017.11.006

Mair, J.; Ritchie, B. W.; Walters, G. Towards a Research Agenda for Post-Disaster and Post-Crisis Recovery Strategies for Tourist Destinations: A Narrative Review. *Curr. Iss. Tour.* **2016,** *19* (1), 1–26. https://doi.org/10.1080/13683500.2014.932758

Mansfeld, Y. Cycles of War, Terror, and Peace: Determinants and Management of Crisis and Recovery of the Israeli Tourism Industry. *J. Travel Res.* **1999,** *38* (1), 30–36. https://doi.org/10.1177/004728759903800107

Mao, Y.; He, J.; Morrison, A. M.; Andres Coca-Stefaniak, J. Effects of Tourism CSR on Employee Psychological Capital in the COVID-19 Crisis: From the Perspective of Conservation of Resources Theory. *Curr. Iss. Tour.* **2020,** 1–19. https://doi.org/10.1080/13683500.2020.1770706

Mtimet, N.; Wanyoike, F.; Rich, K. M.; Baltenweck, I. Zoonotic Diseases and the COVID-19 Pandemic: Economic İmpacts on Somaliland's Livestock Exports to Saudi Arabia. *Global Food Secur.* **2021,** *28*, 100512. https://doi.org/10.1016/j.gfs.2021.100512

Ozbay, G.; Sariisik, M.; Ceylan, V.; Çakmak, M. A Comparative Evaluation Between the İmpact of Previous Outbreaks and COVID-19 on the Tourism İndustry. *Int. Hosp. Rev.* (ahead-of-print) **2021**. https://doi.org/10.1108/IHR-05-2020-0015

Pennington-Gray, L.; Thapa, B.; Kaplanidou, K.; Cahyanto, I.; McLaughlin, E. Crisis Planning and Preparedness in the United States Tourism Industry. *Cornell Hosp. Quart.* **2011,** *52* (3), 312–320. https://doi.org/10.1177/1938965511410866

Perles-Ribes, J. F.; Ramón-Rodríguez, A. B.; Moreno-Izquierdo, L.; Torregrosa Martí, M. T. Winners and losers in the Arab uprisings: A Mediterranean Tourism Perspective. *Curr. Iss Tour.* **2018,** *21* (16), 1810–1829. https://doi.org/10.1080/13683500.2016.1225697

Pizam, A. *Tourism, Security and Safety*. Routledge, 2006.

Renzaho, A. M. N. The Need for the Right Socio-Economic and Cultural Fit in the COVID-19 Response in Sub-Saharan Africa: Examining Demographic, Economic Political, Health, and Socio-Cultural Differentials in COVID-19 Morbidity and Mortality. *Int. J. Environ. Res. Public Health* **2020,** *17* (10). https://doi.org/10.3390/ijerph17103445

Ritchie, B. Tourism Disaster Planning and Management: From Response and Recovery to Reduction and Readiness. *Curr. Iss Tour.* **2008,***11* (4), 315–348. https://doi.org/10.1080/13683500802140372

Ritchie, B. W. Chaos, crises and disasters: A strategic approach to crisis management in the tourism industry. *Tour. Manage.* **2004,** *25* (6), 669–683. https://doi.org/10.1016/j.tourman.2003.09.004

Ritchie, B. W.; Dorrell, H.; Miller, D.; Miller, G. A. Crisis Communication and Recovery for the Tourism Industry. *J. Travel Tour. Market.* **2004,** *15* (2–3), 199–216. https://doi.org/10.1300/J073v15n02_11

Ryu, K.; Bordelon, B. M.; Pearlman, D. M.Destination-Image Recovery Process and Visit Intentions: Lessons Learned from Hurricane Katrina. *J. Hosp. Market. Manage.* **2013,** *22* (2), 183–203. https://doi.org/10.1080/19368623.2011.647264

Santana, G. Crisis Management and Tourism. *J. Travel Tour. Market.* **2004,** *15* (4), 299–321. https://doi.org/10.1300/J073v15n04_05

Selmi, N.; Dornier, R. Perspectives on the Destination İmage of Tunisia. *Worldwide Hosp. Tour. Themes* **2017,** *9* (5), 564–569. https://doi.org/10.1108/WHATT-07-2017-0037

Seyfi, S.; Hall, C. M.; Shabani, B. COVID-19 and İnternational Travel Restrictions: The Geopolitics of Health and Tourism. *Tour. Geogr.* **2020,** 1–17. https://doi.org/10.1080/14616688.2020.1833972

Swaikoski, D. Leisure in the Time of Coronavirus: Indigenous Tourism in Canada and the İmpacts of COVID-19. *World Leisure J.* **2020,** *62* (4), 311–314. https://doi.org/10.1080/16078055.2020.1825266

Thommandru, A.; Espinoza-Maguiña, M.; Ramirez-Asis, E.; Ray, S.; Naved, M.; Guzman-Avalos, M. Role of Tourism and Hospitality Business in Economic Development. *Mater. Today: Proc.*. **2021**. https://doi.org/10.1016/j.matpr.2021.07.059

Vogel, H. L. *Travel İndustry Economics: A Guide for Financial Analysis*; Springer Nature, 2021.

Wang, H.-Y. Determinants Hindering the İntention of Tourists to Visit Disaster-Hit Destinations. *Curr. Iss Tour.* **2017,** *20* (5), 459–479. https://doi.org/10.1080/13683500.2015.1062471

Wen, J.; Wang, W.; Kozak, M.; Liu, X.; Hou, H. Many Brains Are Better Than One: The İmportance of İnterdisciplinary Studies on COVID-19 in and Beyond Tourism. *Tour. Recreat. Res.* **2021,** *46* (2), 310–313. https://doi.org/10.1080/02508281.2020.1761120

Wu, L.; Shimizu, T. Analysing Dynamic Change of Tourism Destination İmage Under the Occurrence of a Natural Disaster: Evidence from Japan. *Curr. Iss Tour.* **2020,** *23* (16), 2042–2058. https://doi.org/10.1080/13683500.2020.1747993

Wut, T. M.; Xu, J.; Wong, S. Crisis Management Research (1985–2020) in the Hospitality and Tourism İndustry: A Review and Research Agenda. *Tour. Manage.* **2021,** *85*, 104307. https://doi.org/10.1016/j.tourman.2021.104307.

Yeh, S.-S. Tourism Recovery Strategy Against COVID-19 Pandemic. *Tour. Recreat. Res.* **2021,** *46* (2), 188–194. https://doi.org/10.1080/02508281.2020.1805933

Zeng, B.; Carter, R. W.; De Lacy, T.öShort-Term Perturbations and Tourism Effects: The Case of SARS in China. *Curr. Iss Tour.* **2005,** *8* (4), 306–322. https://doi.org/10.1080/13683500508668220

CHAPTER 7

Pandemic Management in the Tourism Industry: The Role of Government in Mitigating COVID-19

JEETESH KUMAR[1], FARHAD NAZIR[2], and GÜL ERKOL BAYRAM[3]

[1]*School of Hospitality, Tourism and Events; Centre for Research and Innovation in Tourism (CRiT); Sustainable Tourism Impact Lab, Taylor's University, Malaysia*

[2]*University of Coimbra, CEGOT, Faculty of Arts & Humanities, Portugal; University of Swat, Pakistan*

[3]*School of Tourism and Hospitality Management, Sinop University, Turkey*

ABSTRACT

Recently, pandemic has affected the tourism industry globally. The COVID-19 pandemic has flourished across media and under several titles, including sanitary crisis, health emergency, natural disaster, etc. What outcomes are there to this selective outlining of the situation, and what can it tell us about the nature of the global response? Negative news about political, economic, or social status is distributed worldwide through numerous media channels. Generally, the key to success is creating a problem-solving action plan including prevention measures that combine all stakeholders' interests and activities. However, in the tourism industry, the equal integration of all stakeholders is most complicated. Crisis management, in general, can be described as measures of all types which allow a business to cope with a suddenly occurring danger or risk situation to return as quickly as possible to a regular business routine. Hence,

Dynamics of the Tourism Industry: Post-Pandemic and Post-Disaster Perspectives and Strategies.
Gül Erkol Bayram and Anukrati Sharma (Eds.)

it is necessary to have a systematic and conceptional approach to offer suitable solutions for stakeholders—how tourism businesses react to such crises and what measures should be taken; also, to understand how tourism businesses should prepare themselves for such crises and strategies to conquer them. In this context, the chapter aims to explore the literature on pandemic management in tourism and to identify the motivations of the current academic discourse.

7.1 INTRODUCTION

Tourism contributes significantly to the global economy that accounts for 10.4% (2,9 trillion USD) of global gross domestic product (WTTC, 2019) which is a crucial source of income, jobs, and wealth creation in numerous countries. Tourism has tremendous relevance in numerous aspects of human life: economic, environmental, social, and cultural. Therefore, tourism cannot be seen as a stand-alone industry; its existence elicits infinity multiples in industry and other spheres. Tourism, in short, is turning into one of the foremost profitable growth engines for the worldwide economy. Despite occasional shocks, it sustained growth within the last ten consecutive years from 2009 to 2019 (UNWTO, 2020). Since 1995, the trend of international tourist arrivals has constantly increased, indicating promising tourism growth.

Tourism and pandemic have an unhappy bond predominantly in the postindustrial development milieu. In the twentieth century alone, a list of health pandemics outbreaks prevailed in the length and breadth of planet earth, in addition to the loss of lives, socioeconomic sufferings, and isolation. The positive side of communicational development has been a prominent reason for the fast spread of these pandemics. In similar regard, the travel sector was enacted as a medium for distributing these outbreaks into neighboring and regional entities, further worsening the impacts.

The COVID-19 crisis has authorities worldwide operating in a context of radical uncertainty and faced with difficult trade-offs given the health, economic, and social challenges it raises. Within the first three months of 2020, the novel coronavirus developed into a global pandemic. Schools and universities were closed in spring 2020 for more than one billion students of all ages. COVID-19 spread to almost all countries and affected more than 50 million people worldwide, resulting in more than 1.25 million deaths from November 2019 till the present (Ingravallo, 2020). More than half of the world's population has experienced a lockdown with strong containment measures—the first time in history that such measures are applied on such a large scale.

In sudden, the Coronavirus pandemic has brought exceptional shocks to the travel and tourism industry universally since 2019, as worldwide governments constrain international travel, and even local movements are governed by state legislation. Tourism that works once people travel outside their residency automatically freezes and meets the needs of tourists' sectors both directly and indirectly. United Nations Conference on Trade and Development (UNCTAD) estimated that the loss suffered is 1.5% (1.2 trillion USD) of the global gross domestic product, and further UNCTAD estimates that for each 1 million USD lost in international tourism revenue, a country's value might decline by 2–3 million USD, with international tourist arrivals crashed to 74% less in 2020 which is equivalent to one billion tourists (UNWTO, 2020). UNWTO launched "One billion tourists: One Billion Opportunities" as a tourism campaign, to face this pandemic in 2020, UNWTO released "Restart Tourism" as the newest global travel campaign to stimulate tourism which has the facility to bring individuals back along, provide persistent experiences whereas additionally supporting jobs, serving to businesses and protective culture and natural heritage. The rise of tourism indicates any other business revivals. While considering the above issues, this chapter explores the relevant literature regarding pandemics history, global impacts on the tourism industry, and synching with the best practices adopted by state regimes in mitigating COVID-19 (Saarinen, 2019).

7.2 HISTORICAL CHRONOLOGY OF VIRAL PANDEMICS AND TOURISM

To recognize the pandemic impacts on the tourism industry, it is imperative to have a supervisory glance at the historical flow chart of these happenings in the last hundred years. The illustrative table below is the list of pandemics, their impacts, and steps taken to curb the spread.

Like all pandemics, COVID-19 has a spatial dimension that must be addressed. By November 2020, it will be evident that the COVID-19 crisis has had a significant impact on countries and regions and municipalities within countries in terms of declared cases and deaths. In the People's Republic of China (hereafter referred to as "China"), Hubei province accounted for 83%

TABLE 7.1 Major Health Outbreaks in the Last 100 Years.

Pathogen	Year	Fatality	Impacts on Tourism	Authors
Spanish Flue	1918–1920	50 million	Ban on leisure activities, mass gatherings, travel.	Gallo and Davis (2014); Nascimento (2020)
Asian Flu	1957–1957	02 million	Innovative restrictions on educational premises, travel, and social gatherings.	Henderson et al. (2009); Saunders-Hastings and Krewski (2016)
SARS	2002–2003	774	Decrease of 1.2% in international Tourism Arrival, Cancellation of sports tourism events, cancellation of conferences and business galas.	WHO (2015), Wilder-Smith (2006)
Swine Flu	2009	284,000	Travel advisories, quarantine, thermal screening.	WHO (2010), Page et al. (2012)
MERS	2012	858	Cease of international travel, quarantine.	WHO (2019), Choe et al. (2021)
COVID-19	2019–Present	4,539,723 & counting.	Ban on air, sea, and land travel within and across countries, closure of all public spaces, restaurants, hotels, destinations etc.	WHO (2021)

of confirmed cases. The north of Italy was the hardest hit, with Lombardy, one of Europe's wealthiest areas, recording the highest number of cases (47% as of November) (COVID-19, Italian Government.). New York had the highest percentage of federal cases (14.6%) in the United States, followed by Texas (8%). As of November, the provinces of Quebec and Ontario accounted for 61 and 31% of total cases in Canada, respectively. As of November, metropolitan Santiago accounted for 70% of all cases in Chile. As of November, Sao Paulo, Brazil, had recorded 25% of the cases. In India, Maharashtra accounted for 21% of confirmed cases, while Moscow accounted for 24% of total cases in Russia as of November. COVID-19-related death rates show an intense regional concentration as well. COVID-19 deaths per 100 000 people vary significantly from nation to country, especially in the most hard-hit countries. Calabria, for example, is the least impacted region in Italy, with 5.5 deaths per 100,000 people compared to 171 deaths per 100,000 in Lombardy. In the United States, Vermont had 9.3 fatalities per 100,000 people compared to 184 in New Jersey. In Brazil, Minas Gerais had 41.8 fatalities per 100,000 people, while the mortality rate in the Distrito General was 120. South Korea and New Zealand were less affected than the rest of the world. In Sejong, there were no deaths per 100,000, while in Daegu, there were 8.1 deaths per 100,000 (Ministry of Health of Chile, 2020).

7.3 THE ROLE OF GOVERNMENT IN MITIGATING COVID-19

7.3.1 PREVACCINE MEASURE

In the face of health emergencies, primarily, as evident from the ongoing pandemic, the state fronts have been the focal spheres facing the impacts and devising the management for the mitigation. The unprecedented outcomes of a pandemic have necessitated the state actors to take some of the least expected movement restrictions and lockdowns. Considering the authoritative mandate of regimes, the imposition of mobility control notifications and do and don'ts implementations have been noticeably observed and practised by the populace. However, criticism has been equally echoed on these state restrictions in the context of different countries (Berman, 2020; Geloso and Murtazashvili, 2020; Mendolia et al., 2021; Moser and Yared, 2020; Shaw et al., 2006). The states are in a constant dual push in imposing the restrictions, and in case of easing these. Compromise on social distancing has resulted in economic downfall. In developing and least developed countries, additional socioeconomic exertion has obliged the governments to take unwelcomed decisions (Gronbach and Seekings, 2021).

Contrary to this, in developed countries, two-way communication has been initiated to synchronize the restrictions with the economic needs of residents (Choi et al., 2021). Similarly, the developed regimes have introduced and handed over the economic stimulus packages for the economic sustenance of their citizens (Kaplan et al., 2020). While for some nation-states, bare minimum mobility has been permitted to maintain the operational cycle (Dobusch and Kreissl, 2020).

Hotel and restaurant restrictions have been another contested debate in the academic archives in the hospitality sector alone. Controlled Dining and No-Dining in restaurants and minimum occupational protocols for hotels and periodic operational timings have been devised and implemented by governments across the globe (Bharwani and Mathews, 2021; Brizek et al., 2021; Hoang et al., 2021; Milwood and Crick, 2021; Nhamo et al., 2020;

Singh et al., 2021). On one side, these restraints have deleteriously affected the hospitality business. On the other side, it has obliged the stakeholders to modify their operational models to be more contactless and digital (Byrd et al., 2021; Lai et al., 2020; Sirimongkol, 2021). The travel side being pivotal on mobility has been the leading one in the face of pandemic consequences. All three segments of travel, including the terrestrial, marine, and air, have been equally affected and cancellation and ban imposed on inbound and outbound arrivals.

7.3.2 POSTVACCINE MEASURES

In the contest of vaccination production, several pharmaceutical companies have been consistently competing to launch the vaccine for pandemics as early as possible. Vaccination has been foreseen as another essential entity in reverting the pandemic and a license for a new normal (Ahuja et al., 2021; Rouatbi et al., 2021; Turhan et al., 2021). On the side of supply, there has been a collaborative approach followed by the countries in rolling out the vaccines. Soon after distributing vaccines, countries have revisited their plans and restrictions. To this end, the inaugural of specific travel prerequisites has also been made mandatory for the mobility of travelers. Vaccination Passports, Digital Vaccination Credentials/Certificates, among many others, have been the more pertinent ones to mention. In addition, vaccination proof has been set conditional for dining in restaurants, accommodation in lodging facilities, boarding in terrestrial, marine and air transportation. Regionally, some integrated portals, for example, European Union Digital Certificate, have also been introduced to allow the smooth travel of residing population among member states.

Interestingly, there has been extensive debate made on the pros and cons of these health or immunisation certificates and their acceptability and credibility around the global arena (Corpuz, 2021; Liew and Flaherty, 2021; Mbunge et al., 2021; Satria et al., 2021; Voo et al., 2020). Some riots and protests have been observed against these updated travel and tourism perquisites in some countries. Moreover, excluding some vaccines brands not approved by the European Medicine Agency EMA has also created severe uncertainties for tourists who have already jabbed these nonapproved shots.

7.3.4 LOCKDOWN—MOVEMENT CONTROL ORDER

To reduce the enormous expenses of national incarceration, many governments have implemented localized lockdown measures. This is true, for example, in Aberdeen (Scotland), Auckland (New Zealand), Barcelona (Spain), Melbourne (Australia), various Indian provinces, and some German districts. Such a territorially differentiated approach can avoid the high costs of national confinement while allowing more tailored remedies to problems where they emerge. Policies in federal countries are defined at the state level and hence differ (Castex, 2020). Effective collaboration between local authorities, health agencies, and the federal government is required to handle local outbreaks.

7.3.5 TESTING AND TRACKING

Exit plans from confinement must include testing as a critical component. The WHO advises widespread testing to combat the coronavirus because the pandemic is still in its early stages. Frequent viral testing aids in identifying and isolating infected persons before symptoms appear, reducing the likelihood of second waves. To decrease the danger of future COVID-19 outbreaks, the OECD recommends that 70–90% of all persons who have come into contact with an infected person be tracked, tested, and isolated if infected (OECD, 2020). This necessitates a significant increase in testing, which can be costly. However, the difficulties and expenses of doing so pale in contrast to the costs of lockdowns (OECD, 2020).

The WHO proposes testing by detecting infectious persons and isolating contact cases before symptoms appear (WHO, 2020). Korea's effective strategy in managing the first wave of infections and preventing a second was based on testing and contact tracing, with local governments accountable for COVID-19 screening stations that allowed for rapid and safe testing and monitoring of people in self-quarantine. Between April and October 2020, European nations significantly boosted their capacity and broadened their testing for suspected instances. Official data from the EU27 reveals that over 6 million RT-PCR tests were performed per week in October (ECDC, 2020; Government of the United States of America—White House, 2020).

Subnational governments are at the forefront of putting the "track, isolate, test, and treat" policy into action. While central governments must provide financial resources and coordination in more decentralized situations, regional and local governments will be responsible for policy implementation. Local and regional governments help organize testing and isolation measures in nations with more centralized health service delivery. It is critical to provide an opportunity for local initiatives and experimentation in either a decentralized or a centralized setting. This helps manage the pandemic's asymmetric impact, which frequently necessitates a fast local-level reaction to identify and control clusters.

7.3.6 SPANNING THE EMERGENCY

The COVID-19 epidemic is putting pressure on all levels of government to respond in an environment of enormous uncertainty and economic, budgetary, and social pressure. A new problem has emerged since mid-2020, particularly with the start of the second wave of infections in many countries: the restricted ability to sequence policy action. To manage, escape, and recover from the crisis, national, regional, and municipal governments have discovered that they cannot rely on a straight or linear route of policy action. Governments must instead act in lockstep on all fronts simultaneously (SKYtg24 , 2020).

Governments are rethinking their multi-level governance structures, revaluating their policy instruments, and rethinking their regional development goals in response to this requirement for flexibility and adaptation. The adoption of a place-based strategy and mobilizing and coordinating different policy sectors and all levels of government are required for success. It requires strong leadership and sound coordination, consultation, and collaboration between government and nongovernment players.

It also depends on regaining public trust and making the most of continuing engagement with stakeholders and people (CoR-OECD, 2020). The responses to COVID-19 show promise in terms of regional development goals, with more accessible essential services, fewer digital divides, changed global value chains and industrial policies, and more excellent climate action, among other things, is highlighted.

7.3.7 EFFECTIVE GOVERNMENT

In this epidemic, effective government public communication is critical to ensure that the official message is consistent internally and with the public and civil society. Behavioral communication initiatives have aided in implementing rules by encouraging or teaching large parts of the community to comply with needed procedures, such as hand washing and observing lockdown and social separation restrictions (Government of United Kingdom, 2020).

Citizens' faith in and participation in government policy may be strengthened via effective communication. In the battle against deception and misinformation, it is critical. Finally, it may assist in reaching out to specific demographic groups and facilitating discussion with individuals to ensure that policies and services are tailored to their needs and expectations.

7.3.8 POST-COVID-19 DEVELOPMENT

COVID-19's differing effects on individuals, communities, and regions and the possibility of exacerbating territorial imbalances provide a place-based approach to regional development and greater inclusivity a new urgency. The necessity of a balance between top-down and bottom-up activities, the need for successful relationships and trust across various actors, flexibility and adaptation, and the value of a balance between top-down and bottom-up actions contribute to this urgency. It's also reignited policy debates on regional resilience. As a result of the epidemic and the pressures it throws on all levels of government, regional development objectives are shifting to bolstering regional resilience (OECD Trento Centre for Local development, 2020). To create more resilient regions, first, ensure that they can absorb, recover (or bounce back) from, and/or adapt to the impact of economic, financial, environmental, political, and social shocks or chronic pressure; and then ensure that they can continue to meet the needs of citizens and businesses at least as well as—and ideally better than—before the crisis. Building more resilient regions after COVID-19 may increase national and subnational investment in health care and other public services.

In the EU, 76% of regional and municipal governments questioned agree that regional development policies should prioritize access to high-quality public services, including health, across all regions. It is also likely to lead to a re-evaluation of regional policy objectives, including their urban/rural balance, the digital divide, the balance between tangible and intangible assets (infrastructure such as broadband, public transportation, and social housing), R&D, innovation, well-being, and culture, productivity, and industrial profiles, and how to best meet higher-level aims, in addition to other issues.

In a crisis, the most typical focus of government centers is on managing and coordinating government operations. According to the OECD's 2017 Survey on the Organisation and Functions of the Centre of Government, 83% of CoGs have some risk management responsibilities, with over a third having primary responsibility. Despite these numbers, only around 10% of CoGs identified "risk management and strategic foresight for the whole government" as a critical role of the Centre (OECD, 2018). The government's reaction to the coronavirus (COVID-19) pandemic illustrates how it adapts its crisis management to a complex crisis and institutional context. Governments have frequently created supplementary ways to standard emergency management processes, which are generally led or backed by the center of authority. They had no choice since they were up against a slew of "unknown unknowns."

7.3.9 TRAVEL RESTRICTIONS

The majority of the UNWTO Panel of tourism impacts of COVID-19 estimate international tourists may only be bound in 2024 or later. To accelerate the rebound of tourism, the role of the Government in Mitigating COVID–19 is very crucial, not only to keep people safe but also to keep businesses functional, as suggested by UNWTO as follows; build recovery strategies, speed up the pace of worldwide vaccination, functioning on effective coordination and communication on dynamical travel restrictions whereas advancing digital tools to facilitate quality are going to be crucial in building tourists' trust to restart tourism as fast as possible by having best global health and prudent guidelines so that tourists can start arriving (Devi, 2020). Restarting tourism requires many adjustments of health infrastructure for tourism destinations and businesses, which costs money. Grants and low-cost loans from regional and international financial institutions can offer the necessary resources to make such changes and increase visitor appeal. Small, local businesses should be the focus of such programs. Committee for Economic and Commercial Cooperation (COMCEC) suggests more than just financial aids but also stimulating demand aids, legal aids, and operational aids.

One of the critical steps to keep tourism alive during this pandemic is adapting to a changing environment by applying travel restrictions. As a result of the pandemic, 96% of all worldwide destinations have imposed travel restrictions since April 2020, where the restrictions given are also continuously updated with new virus variants. In most cases, one mandatory travel restriction is in the form of vaccination requirements (tourists completed two doses of associate approved vaccine) and health protocol application during stays (wear a double mask, maintain physical distance, and clean hand frequently); other restrictions are according to their respective destinations (Vaidya et al., 2020).

7.3.10 *TRAVEL BUBBLE*

One more step to stimulate tourism as quickly as possible is called a travel bubble. A travel bubble is a travel corridor between countries amid a pandemic based on neighboring nations' agreement that has successfully controlled the COVID-19 to allow community travels across their borders only. Hence, the community is free to move about within the bubble, but they cannot enter outside (Yu et al., 2020). The goal is to provide more freedom without inflicting more harm. One of the advantages of the travel bubble approach is that the measurements used in each bubble may be modified to equalize infection risk efficiently and risk levels across origin and destination nations and empower businesses to restart. There are three main types of travel bubbles: The Basic Travel Bubble (BTB), which includes a basic set of standards, The Limited Travel Bubble (LTB) with additional test within 24–48 after departure, The Extended Travel Bubble (ETB) with two additional tests; 24–48 h after departure and 24–48 h after the arrival.

7.4 CONCLUSIONS

COVID-19, which seriously harms human health, has spread to a large part of the world in a concise time and has put countries in difficulties in health and many areas. Countries have tried to save both their public health and their economies with their measures and the bans they have implemented. One of the most sectors by the COVID-19 outbreak is the tourism sector. Tourism has been affected by COVID-19 in two different ways. The first effect is the inability to carry out tourism due to travel restrictions and quarantines. The second effect is that individuals in economic distress cannot participate in tourism activities.

Both reasons have brought tourism activities to a standstill. According to the World Tourism Organization (UNWTO) reports, there has been a rapid decline in tourist arrival in COVID-19. According to Dwyer et al. (2006), COVID-19 caused a much sharper decline than the previous crises, such as the 2008 economic crisis, the Iraq War, and the SARS virus. The primary way to get out of the crisis with minimum damage is the directing and protective tourism policies that the governments will take. When the measures for tourism and the national economy of the countries in the top ten in international tourist arrivals are examined, it is seen that applications are made in favor of citizens and businesses by giving up. However, in this process, countries reduced taxes or did not take them entirely temporarily, helping to allocate and amount of loan interest rates, and postponing the overdue debts. The personnel were exempted from other taxes by partially paying their salaries and supporting their insurance expenses to reduce employment. It has secured its expenses with cash assistance to those in need.

According to Evans (2020), the common points of the measures taken by other countries such as England, Australia, Greece, Malaysia, and the Netherlands, which are in the first place in tourist arrivals; partial payment of employee salaries by the state, tax exemptions, granting vouchers with state guarantee instead of cancellation of reservations, increasing the rate of state guarantee applied on loans, allocating funds to cover cash expenses, and not applying taxes in different categories in 2020. States' policies and plans in times of crisis have made this process healthier and more resilient (Gopinath, 2020).

As it has been explained in the chapter, the process related to COVID-19 continues, and the necessity of taking the activities to be carried out, the measures to be taken, and the plans to be implemented in this axis should be revealed. In this context, a more advanced inspection of hygiene and distance rules in every layer of the tourism sector will sustainably increase tourist mobility. However, extending the economic support packages of the states to a longer term will significantly reduce the effects of the current crisis. The resilience of destinations and their capacity to crises is a necessary and needed issue for sustainable tourism. The implementation of the standards set by institutions such as UNWTO and WWTC, together with the individual measures of the countries, can reduce the spread of the virus and accelerate the tourist flow.

KEYWORDS

- **COVID-19**
- **pandemic**
- **tourism industry**
- **role of government**
- **vaccine**
- **movement control order**

REFERENCES

Ahuja, A.; Athey, S.; Baker, A.; Budish, E.; Castillo, J. C.; Glennerster, R.; Kominers, S. D.; Kremer, M.; Lee, J.; Prendergast, C.; Snyder, C. M.; Tabarrok, A.; Tan, B. J.; Więcek, W. Preparing for a Pandemic: Accelerating Vaccine Availability. *AEA Papers Proc.* **2021,** *111*, 331–335. https://doi.org/10.1257/pandp.20211103

Berman, E. The Roles of the State and Federal Governments in a Pandemic. *J. Natl. Secur. Law Policy* **2020,** *11*, 61.

Bharwani, S.; Mathews, D. Post-Pandemic Pressures to Pivot: Tech Transformations in Luxury Hotels. *Worldwide Hosp. Tour. Themes* (ahead-of-print). **2021**. https://doi.org/10.1108/WHATT-05-2021-0072

Brizek, M. G.; Frash, R. E.; McLeod, B. M.; Patience, M. O. Independent Restaurant Operator Perspectives in the Wake of the COVID-19 Pandemic. *Int. J. Hosp. Manage.* **2021,** *93*, 102766. https://doi.org/10.1016/j.ijhm.2020.102766

Byrd, K.; Her, E., Fan, A.; Almanza, B.; Liu, Y.; Leitch, S. Restaurants and COVID-19: What Are Consumers' Risk Perceptions About Restaurant Food and İts Packaging During the Pandemic? *Int. J. Hosp. Manage.* **2021,** *94*, 102821. https://doi.org/10.1016/j.ijhm.2020.102821

Castex, J. *Plan de preparatıon de la sortıe du confinement. Paris : Direction de l'information légale et administrative.* 2020, https://doc.santelysformation.fr/index.php?lvl=notice_display&id=60554. (Accessed August 23 2021.)

Choe, Y.; Wang, J.; Song, H. The impact of the Middle East Respiratory Syndrome Coronavirus on İnbound Tourism in South Korea Toward Sustainable Tourism. *J. Sustain. Tour.* **2021,** *29* (7), 1117–1133. https://doi.org/10.1080/09669582.2020.1797057

Choi, K.; Giridharan, N.; Cartmell, A.; Lum, D.; Signal, L.; Puloka, V.; Crossin, R.; Gray, L.; Davies, C.; Baker, M. Life During Lockdown: A Qualitative Study of Low-İncome New Zealanders' Experience During the COVID-19 Pandemic. *New Zealand Med. J.* **2021,** *134* (1538), 52–56.

Corpuz, J. C. G. COVID-19 Vaccination Certificate (CVC) for ASEAN: The Way Forward? *J. Public Health* **2021**, *fdab202*. https://doi.org/10.1093/pubmed/fdab202

CoR-OECD. *CoR-OECD Survey Questionnaire: "How İs COVID-19 Affecting Regions and Cities?*, 2020. https://cor.europa.eu/en/news/Pages/ECON-cor-oecd-survey-COVID-19.aspx. 2020, (Accessed June 13 2021).

Devi S. Travel Restrictions Hampering COVID-19 Response. *Lancet (London, England)* **2020,** *395* (10233), 1331–1332. https://doi.org/10.1016/S0140-6736 (20)30967-3.

Dobusch, L.; Kreissl, K. Privilege and Burden of İm-/Mobility Governance: On the Reinforcement of İnequalities During a Pandemic Lockdown. *Gender Work Organ.* **2020,** *27* (5), 709–716. https://doi.org/10.1111/gwao.12462

Dwyer, L.; Forsyth, P.; Spurr, R.; Van Ho, T. Economic Effects of the World Tourism Crisis on Australia. *Tour. Econ.* **2006,** *12* (2), 171–186. DOİ: 10.5367/000000006777637467.

ECDC *COVID-19 Data Plateform*. https://www.ecdc.europa.eu/en/COVID-19/data. 2020 (Accessed 25 Aug 2021).

Gallo, M. I. P.; Davis, R. A.*The Spanish influenza Pandemic of 1918–1919: Perspectives from the Iberian Peninsula and the Americas Boydell & Brewer*, Vol. 30, 2014.

Evans, O. Socio-Economic İmpacts of Novel Coronavirus: The Policy Solutions. *Biz Econs Quart.* **2020,** *7*, 3-12.

Geloso, V.; Murtazashvili, I. Can Governments Deal with Pandemics? 2020. Available at SSRN 3671634.

Government of United Kingdom. *Government Cracks Down on Spread of False Coronavirus İnformation Online*, 2020. https://www.gov.uk/government/news/government-cracks-down-on-spread-of-false-coronavirus-information-online.2020 (Accessed 25 Aug 2021).

Government of the United States of America—White House *Vice President Pence and Secretary Azar Add Key Administration Officials to the Coronavirus Task Force*. The White House, 2020. https://www.whitehouse.gov/briefings-statements/vice-president-pence-secretary-azar-add-key-administration-officials-coronavirus-task-force-2/ (Accessed 25 Aug 2021).

Gopinath, G. Limiting the Economic Fall Out of the Coronavirus with Large Targeted Policies. Mitigatingthe COVID Economic Crisis: Act Fastand Do Whatever It Takes, 2020; pp 41–48

Gronbach, L.; Seekings, J. Pandemic, Lockdown and the Stalled Urbanisation of Welfare Regimes in Southern Africa. *Global Soc. Policy* **2021,** 14680181211013724. https://doi.org/10.1177/14680181211013725.

Henderson, D. A.; Courtney, B.; Inglesby, T. V.; Toner, E.; Nuzzo, J. B. Public Health and Medical Responses to the 1957-58 Influenza Pandemic. *Biosecur. Bioterrorism* **2009,** *7* (3), 265–273. https://doi.org/10.1089/bsp.2009.0729

Hoang, T. G.; Truong, N. T.; Nguyen, T. M. The Survival of Hotels During the COVID-19 Pandemic: A Critical Case Study in Vietnam. *Serv. Busi.* **2021,** *15* (2), 209–229. https://doi.org/10.1007/s11628-021-00441-0

Kaplan, G.; Moll, B.; Violante, G. L. *The Great Lockdown and the Big Stimulus: Tracing the Pandemic Possibility Frontier for the US*; National Bureau of Economic Research, 2020.

Ingravallo, F. Correspondence Death in the Era of the. *Lancet Public Health* **2020,** *5* (5), e258.

Lai, H. B. J.; Abidin, M. R. Z.; Hasni, M. Z.; Ab Karim, M. S.; Ishak, F. A. C. Key Adaptations of SME Restaurants in Malaysia Amidst the COVID-19 Pandemic. *Int. J. Res. Busi. Soc. Sci.* **2020,** (2147–4478), *9* (6), 12–23.

Liew, C. H.; Flaherty, G. T. Immunity Passports to Travel During the COVID-19 Pandemic: Controversies and Public Health Risks. *J. Public Health* **2021,** *43* (1), e135–e136. https://doi.org/10.1093/pubmed/fdaa125

Mbunge, E.; Dzinamarira, T.; Fashoto, S. G.; Batani, J. Emerging Technologies and COVID-19 Digital Vaccination Certificates and Passports. *Public Health Practice (Oxford, England)* **2021,** *2*, 100136–100136. https://doi.org/10.1016/j.puhip.2021.100136

Mendolia, S.; Stavrunova, O.; Yerokhin, O. Determinants of the Community Mobility During the COVID-19 Epidemic: The Role of Government Regulations and İnformation. *J. Econ. Behav. Organ.* **2021,** *184*, 199–231. https://doi.org/10.1016/j.jebo.2021.01.023

Milwood, P. A.; Crick, A. P. Culinary Tourism and Post-Pandemic Travel: Ecosystem Responses to an External Shock. *J. Tour. Heritage Serv. Market. (JTHSM)* . **2021,** *7* (1), 23–32.

Ministry of Health of Chile. *Plan de acción Coronavirus*, 2020. https://www.minsal.cl/presidente-pinera-se-reune-con-consejo-asesor-del-minsal-por-COVID-19/ (Accessed 12 Sept 2021).

Moser, C. A.; Yared, P. *Pandemic Lockdown: The Role of Government Commitment*; National Bureau of Economic Research, 2020. DOI: 10.3386/w27062.

Nascimento, A. F. do. The Tourist Topicality of the Case of the Spanish Flu in the City of Rio de Janeiro (Sep. 1918–Mar. 1919). *Revista Brasileira de Pesquisa Em Turismo* **2020,** *14*, 176–188.

Nhamo, G.; Dube, K.; Chikodzi, D. Restaurants and COVID-19: A Focus on Sustainability and Recovery Pathways. In *Counting the Cost of COVID-19 on the Global Tourism Industry*; Nhamo, G., Dube, K., Chikodzi, D., Eds.; Springer International Publishing, 2020; pp 205–224. https://doi.org/10.1007/978-3-030-56231-1_9; 2020.

OECD Trento Centre for Local development/ *Coronavirus (COVID-19): SME Policy Responses—A Case Study: Italian Regions (as of 25 March 2020)*. 2020. https://www.oecd.org/cfe/leed/COVID-19-Italian-regions-SME-policy-responses.pdf (Accessed 16 Sept 2021).

OECD. Centre Stage 2 The Organisation and Functions of the Centre of Government in OECD Countries, 2018. https://www.oecd.org/gov/centre-stage-2.pdf (Accessed 20 Sept 2021).

Page, S.; Song, H.; Wu, D. C. Assessing the Impacts of the Global Economic Crisis and Swine Flu on Inbound Tourism Demand in the United Kingdom. *J. Travel Res.* **2012,** *51* (2), 142–153. https://doi.org/10.1177/0047287511400754

Rouatbi, W.; Demir, E.; Kizys, R.; Zaremba, A. Immunising Markets Against the Pandemic: COVID-19 Vaccinations and Stock Volatility Around the World. *Int. Rev. Financial Analy.* **2021,** *77*, 101819. https://doi.org/10.1016/j.irfa.2021.101819

Satria, F. B.; Khalifa, M.; Rabrenovic, M.; Iqbal, U. Can Digital Vaccine Passports Potentially Bring Life Back to "True-Normal"? *Comput. Methods Programs Biomed.* **2021,** *1*, 100011.

Saunders-Hastings, P. R.; Krewski, D. Reviewing the History of Pandemic Influenza: Understanding Patterns of Emergence and Transmission. *Pathogens (Basel, Switzerland)* **2016,** *5* (4), 66. https://doi.org/10.3390/pathogens5040066

Saarinen, J. What Are Wilderness Areas For? Tourism and Political Ecologies of Wilderness Uses and Management in the Anthropocene. *J. Sustain. Tour.* **2019,** *27* (4), 472-487.

Shaw, K. A.; Chilcott, A.; Hansen, E.; Winzenberg, T. The GP's Response to Pandemic İnfluenza: A Qualitative Study. *Family Practice* **2006,** *23* (3), 267–272. https://doi.org/10.1093/fampra/cml014

Singh, N.; Bhatia, S.; Nigam, S. Perceived Vulnerability of Job Loss and Satisfaction with Life in the Hospitality Sector in Times of Pandemic: A Multi-Mediational Approach. *Int. J. Contemp. Hosp. Manage.* **2021,** *33* (5), 1768–1788. https://doi.org/10.1108/IJCHM-10-2020-1145

Sirimongkol, T. The Effects of Restaurant Service Quality on Revisit İntention in Pandemic Conditions: An Empirical Study from Khonkaen, Thailand. *J. Foodserv. Busi. Res.* **2021,** 1–19. https://doi.org/10.1080/15378020.2021.1941560

SKYtg24. *Chi è Angelo Borrelli, capo della protezione civile nominato commissario per l'emergenza coronavirus*, 2020. https://tg24.sky.it/politica/approfondimenti/angelo-borrelli-chi-e (Accessed 20 Sept 2021).

Turhan, Z.; Dilcen, H. Y.; Dolu, İ. The Mediating Role of Health Literacy on the Relationship Between Health Care System Distrust and Vaccine Hesitancy During COVID-19 Pandemic. *Curr. Psychol.* **2021**. https://doi.org/10.1007/s12144-021-02105-8

UNWTO. UNWTO World Tourism Barometer 2020, *18* (2), May. https://www.e-unwto.org/toc/wtobarometereng/18/2 (Accessed 20 Sept 2021).

Vaidya, R.; Herten-Crabb, A.; Spencer, J.; Moon, S.; Lillywhite, L. Travel restrictions and infectious disease outbreaks. *Journal of travel medicine* **2020,***27* (3), taaa050

Voo, T. C.; Clapham, H.; Tam, C. C. Ethical Implementation of Immunity Passports During the COVID-19 Pandemic. *J. Infect. Dis.* **2020,** *222* (5), 715–718. https://doi.org/10.1093/infdis/jiaa352

Wilder-Smith, A. The Severe Acute Respiratory Syndrome: Impact on Travel and Tourism. *Travel Med. Infect. Dis.* **2006,** *4* (2), 53–60. https://doi.org/10.1016/j.tmaid.2005.04.004
World Travel and Tourism Council (WTTC). 2019. https://www.wttc.org/economic-impact/ (2018) (Accessed August 30 2021).
World Health Organization. *World Health Statistics 2015*; World Health Organization, 2015.
World Health Organization. *World Health Statistics 2010*; World Health Organization, 2010.
World Health Organization. *World Health Statistics 2019*; World Health Organization, 2019.
World Health Organization. *World Health Statistics 2019*; World Health Organization, 2021.
Yu, J. H.; Lin, H. H.; Lo, Y. C.; Tseng, K. C.; Hsu, C. H. Is the Travel Bubble Under COVID-19 a Feasible Idea or Not? *Int. J. Environ. Res. Public Health* **2021,** *18* (11), 5717.

CHAPTER 8

New Normal Policies, Regulations, and Strategies Related to the Tourism Industry

HASAN TAHSIN KAVLAK and SEDA ERKEKLI

Faculty of Tourism, Recreation Management, Ankara Hacı Bayram Veli University, Ankara, Turkey

ABSTRACT

In this study, information is given about new normal policies, regulations and strategies in tourism industry and the importance of viewpoint of a new tourism. Tourism industry is one of the most fragile and flexible sectors in economy. In addition to this, the tourism industry is a social integration because tourism means person. The tourism industry which is a fragile and flexible sector, needs new strategies to deal with unexpected situations. Especially in the COVID-19 epidemic process, the negative effects of epidemics on tourism industry can be decreased with the new tourism strategies and regulations. The adaptation to new strategies usually provides competitive advantage in sector. At the same time, new products such as virtual travel in tourism arise with the new strategies. It is very difficult both to protect people from epidemics and to develop the tourism sector. The aim of the study is to reveal determination of opportunities and threats, new normal policies, strategies, and regulation in the epidemic process.

8.1 INTRODUCTION

Tourism industry: It shows a constant change due to its flexible, fragile, and dynamic structure. While this change in the tourism industry is sometimes

Dynamics of the Tourism Industry: Post-Pandemic and Post-Disaster Perspectives and Strategies.
Gül Erkol Bayram and Anukrati Sharma (Eds.)

caused by the negative effects of changing environmental conditions, sometimes it is the result of adapting to changing environmental conditions. Therefore, changing environmental conditions can be an advantage as well as a disadvantage for the tourism industry. Tourism dynamics that adapt the fastest to environmental conditions can take advantage of change, while tourism dynamics that cannot adapt to environmental conditions may experience the disadvantage of change.

Tourist behavior is affected not only by tourism businesses, but also by changes in the environment and conditions. In particular, force majeure seen in a destination such as epidemics, disasters, wars can directly affect the views of tourists about that destination. However, in cases where such a force majeure affects the whole world, it can be said that the need for tourism is felt more than tourist preferences. With the COVID-19 epidemic, this situation has been experienced recently and continues to be experienced.

In the process of COVID-19, which emerged in Wuhan, China in 2019 and changed the order of the whole world, state governments tried to prevent the epidemic from increasing with some precautions and restrictions. At this point, travel ban, curfews, applications for closing places where people are concentrated, such as restaurants, accommodation establishments, entertainment establishments, shopping malls, are among these restrictions and measures. Therefore, these practices have had a greater impact on the tourism industry. Although people comply with the rules in order not to endanger their health, the prolongation of the epidemic process has caused the needs such as entertainment, socialization, and travel to be felt more intensely. On the other hand, states and businesses also needed tourism mobility. As a matter of fact, since the tourism industry is an economic and social industry on both micro and macro scales, it also provides an added value to the states. Therefore, tourism mobility is also an important industry for countries (Ivanov and Webster, 2007).

All these developments have led both people and states to a way of life called the "new normal." In this new normal process, there have been serious changes in the tourism understanding of tourists, the tourism products offered by the enterprises, and the tourism policies of the states. In this context, this section aims to examine the impact of the COVID-19 epidemic, which is a crisis period for the tourism industry, on tourism and the new normal process in the tourism industry. In line with this purpose, new tourism trends, tourism policies, and tourism understanding developed in the new normal process are discussed.

8.2 EPIDEMICS AND THEIR EFFECTS ON THE TOURISM INDUSTRY

The earth, which has existed for more than 4.5 billion years, has encountered many epidemic diseases such as SARS, Ebola, Spanish flu, cholera, smallpox, avian flu, swine flu, and yellow fever. In the first quarter of the 21st century, the COVID-19 outbreak is the newest known outbreak. Considering that each epidemic environment brings a new order, it can be said that a new social and economic order has emerged with the COVID-19 epidemic. The tourism industry is one of the industries most rapidly affected by this new order, because the tourism sector is one of the sectors with the highest human mobility (Hall, 2015).

The SARS (Severe Acute Respiratory Syndrome) epidemic, which heavily affected Asian tourism in 2003, caused the loss of 3 million tourism workers in China, Hong Kong, Singapore, and Vietnam destinations. In addition, the gross domestic product of these destinations suffered a loss of more than 20 billion dollars and tourism activity in these destinations decreased by 70% (McKercher and Chon, 2004). The first case of the SARS epidemic, which emerged on November 16, 2002, was first seen in China, like COVID-19. Later, it spread to many parts of the world in the form of Hong Kong, Vietnam, Singapore, Canada, Taiwan, Philippines, Australia, and America. Although the first case was seen in November 2002, the first travel warnings were made in April 2003 (Whaley and Mansoor, 2006: 3–47).

Ebola is one of the effective epidemics on the earth, such as the SARS epidemic. The Ebola epidemic, which emerged in the African continent in 1976 and increased its effect by mutating in 2014, caused the death of one out of every two people who contracted the disease. It is stated that as of 2021, cases continue in Guinea and the Democratic Republic of the Congo (WHO, 2021). Due to the Ebola epidemic, 50% of tour operators in Africa cancelled their package tours, while 69% of them reduced their reservations for the coming years. During the epidemic, the Tanzania Hotel Association stated that accommodation decreased by 30 and 40% compared to the previous year (Maphanga and Henama, 2019).

Another epidemic in the 21st century is swine flu. The economic costs of the swine flu epidemic, which first appeared in Mexico in 2009, were low in the short term. But these costs can be much higher in the long term (OEF, 2009). The effect of the swine flu epidemic on tourism was also felt very intensely. The UK, in particular, suffered a loss of 1.6 million visitors and a loss of £940 million in revenue. On the other hand, it is estimated that

19.6% of this loss caused by swine flu belongs to England (Page et al., 2012). Although the UK has a strong economy and tourism potential, it has been affected by swine flu. In addition, destinations such as Brunei, which has a smaller and fragile economy compared to the United Kingdom, were also affected by the swine flu. The destination of Brunei, which is located in a small area in the Southeast Asian continent, also lost 30 000 tourists and 15 million dollars with the swine flu (Haque and Haque, 2018).

One of the epidemics that negatively affect tourism mobility is avian flu. Although avian flu is an epidemic with a high mortality rate, its effects on tourism have not been as intense as other epidemics. Based on the data of 18 countries, the World Health Organization stated that from 2003 to 2021, 456 cases out of a total of 863 were fatal (WHO, 2021). Although the death rate is high, the impact of avian flu on tourism is relatively lower than other epidemics due to the low case rates. Kuo et al. (2008), Lee and Chen (2011) revealed that this effect is relatively low. Lee and Chen (2011) stated that even the elderly Asian individuals, who are most likely to be affected by the epidemic, continue to want to participate in tourism mobility.

Each of these epidemics has had an intense or less frequent impact on tourism, because tourism has a flexible structure and is a sector where human mobility is intense. Therefore, every epidemic period is a crisis period for the tourism sector. However, the crisis experienced in each epidemic period has created a more intense demand in the postepidemic period. Karabulut et al. (2020), based on the tourism data of 129 countries between 1996 and 2018, revealed that the epidemic affected tourism negatively. However, it has also been proven that tourism activity in the postepidemic period is more intense than before the epidemic (Ünlüönen and Çeti, 2019). At this point, although the epidemic creates a crisis environment for a certain period in the tourism sector, it can turn into an opportunity later on. The most effective method in turning these crises into opportunities is to create the right policy and planning.

One of the biggest epidemics in the first quarter of the 21st century is the COVID-19 epidemic. This epidemic, which emerged in 2019, still continues to affect the world. In 2020, there was a period of intense curfews and travel restrictions. Therefore, the tourism sector has also been greatly affected by these restrictions. In the COVID-19 epidemic, which is a crisis for the tourism sector, a policy called the "new normal" has emerged. With the new normal policy, the tourism sector has been reshaped. This change is not yet complete. However, it is possible to say that the destinations that adapt to the new normal process the fastest increase tourism mobility relatively.

8.3 NEW NORMAL POLICIES AND NEW NORMALIZATION IN THE TOURISM INDUSTRY

The tourism sector, being one of the leading sectors in the world, offers employment opportunities to many people. The COVID-19 crisis is considered an opportunity to consider the tourism industry from different perspectives and to design a sustainable future. It is important to establish policies to support travel and tourism activities around the world, to reduce the effects of the COVID-19 crisis on the sector, and to revive tourism. In this context, it is critical to pursue more comprehensive, sustainable policies.

It is necessary to have detailed information about the effects of the crisis to minimize the impact of the crisis that occurred as a result of the COVID-19 epidemic on the tourism sector and economy and to determine the right policies (Skare et al., 2020: 13).

The concept of health comes to the fore in the policies implemented within the scope of the COVID-19 epidemic. The policies implemented as a result of the crisis in the tourism sector with the epidemic differ according to the size of the enterprises. Financial support and tax policies implemented in the tourism sector contribute to the revival of the sector (Çakır and Barakazı, 2020).

In order for the policies to be implemented to reopen and recover the tourism sector to be effective, "the government, development partners and international financial institutions must act in coordination (UNWTO, 2020).

With the epidemic process, the sensitivity on nature has increased. In this direction, it is predicted that "mountain, nature, heritage, culture and adventure" tourism will grow rapidly in the coming years. Within the scope of sustainable tourism, it is important to invest in protected areas for ecotourism, support local groups, combat biodiversity, and protect the ecosystem. Priority will be given to investments that are far from traditional, innovative and have the sustainability theme at the forefront. The concept of sustainability will become an important part of the tourism industry (UNWTO, 2020).

8.4 TOURISM TRENDS IN THE NEW NORMAL PROCESS

With the COVID-19 epidemic that has occurred around the world, there have been some changes in the tourism choices of tourists. These changes have created different types of tourism, and environments where individuality is at the forefront and intertwined with nature have laid the groundwork for an increase in interest in eco-tourism activities. Regions that are easily accessible by personal vehicles have become a center of attraction during

TABLE 8.1 Tourism Policies for the COVID-19 Period.

Policies	Implementations
Financial Policies	Tax exemption/deferral
	Economic funds for SMEs
	investment programs
	Social protection (for the vulnerable and disabled)
	Protection fee related to tourism-related activities to businesses in the travel, hospitality, and other sectors
	Discount on utilities expenses for businesses
	Cash flow assistance to businesses
Monetary Policies	Lowering interest rates
	Defining custom credit limits
	Accessible loan opportunities for SMEs
	Incentive programs for airlines
Business Policies	Lowering education taxes
	Financial support for trainees
	Providing training programs for digital development
	Employment support
	Capacity-building programs
Market Policies	Using digital platforms to inform stakeholders in tourism
	Setting guidelines for businesses
	Incentive initiatives for virtual tourism
Public-Private Sector Partnerships	Private sector solidarity funds
	Business investment supports
	Supporting digitization
Restarting Tourism	Booking flexibility in travel packages
	Health and safety protocols, certifications, and business labels
	Creation of international security corridors
Promoting Domestic Tourism	Domestic travel vacation coupons
	Domestic promotion and marketing activities
	Product development for domestic tourism

Source: Adapted from UNWTO (2021).

the epidemic process. Cultural travels, where individuals will gain different experiences, come to the fore in this process (TÜRSAB, 2020: 22).

The prominent elements in the future tourism trends after the COVID-19 crisis are explained as follows (Lanioglo and Rissanen, 2020: 526–527):

- ***Digitalization:*** It is expected to intensify the use of technology within tourism activities, to provide contactless customer experience, and to provide tourism services for Y and Z generations.
- ***Responsible and Sustainable Tourism:*** With the restrictions applied during the epidemic process, the natural environment has started to renew itself. Awareness of the value of nature constitutes the key point of the concept of sustainable tourism.
- ***Domestic Tourism:*** After the epidemic, individuals will prefer places close to the region they live in order to feel safer. In addition, domestic tourism activities will bring regional development.
- ***Health Services:*** The importance of the concept of health has increased with the epidemic. In this direction, it is expected that travels where health and well-being are at the forefront will be preferred.

In the postepidemic period, individuals will tend to travel to calm areas where they are in control, rather than crowded destinations. In this direction, rural areas intertwined with nature will come to the fore. Accommodation areas such as bungalows, tents, camp-caravans, and villas will be among the preferences that attract great attention. In addition, it is predicted that alternative tourism types such as "health tourism, cultural heritage tourism, avian watching, underwater diving tourism" will be preferred by individuals (Düzgün, 2021: 148).

Technological developments create opportunities for the tourism sector. Technological developments are of special importance to follow innovative strategies within the sector and ensure the success of the business. The concept of technology, which has gained a different dimension in the lives of individuals with the epidemic process, is increasing its importance day by day. Many virtual activities become the focus of attention of individuals. Virtual tours, online seminars, and meetings have come to the fore to evaluate the time spent at home as a result of the restrictions. With these developments, the concept of "e-tour" has emerged by bringing a new dimension to tourism. It is predicted that the interest in e-tours will continue in the future (TUYED, 2021).

It is important to keep up with the changes to gain a competitive advantage in the tourism sector. The concept of digitalization comes to the fore at the point of catching new trends in the sector. Virtual experiences that will enable individuals to explore different worlds, get to know new cultures, and revitalize themselves constitute an important part of the digital world.

The tourism sector is one of the leading sectors that can be digitized at a global level, e.g., flight, reservation, itineraries etc. In applications such as digitalization, the concept of digitalization comes to the fore. Incorporating "Industry 4.0, artificial intelligence, augmented reality" technologies in the sector at the point of adapting to the changes and developments that occur will make the tourism sector more interesting (Topal, 2020).

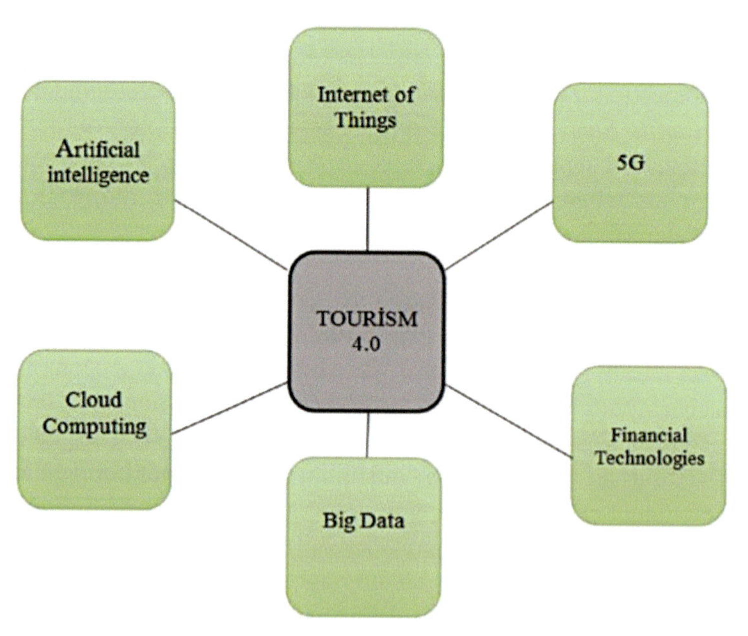

FIGURE 8.1 Tourism 4.0 and potential technologies where it can be interactive.
Source: Reprinted with permission from Callı (2021: 99).

The application of technological developments such as "internet of things, artificial intelligence, virtual and augmented reality, block chain" offered by Industry 4.0 in the tourism sector is seen as an important competitive tool. The use of these applications will add a new dimension to the tourism sector and it will become important to obtain a good visitor experience (Çallı, 2021: 99).

With the epidemic process, the concept of digitalization has settled in the center of both individual and social lives. Considering that digitalization will increase gradually, it is of great importance to redesign the virtual platforms, software, and projects related to this in a way that can meet the wishes and needs of individuals. In the tourism sector, virtual reality tools can be applied as innovative strategies to provide individuals with multiple emotional experiences, customer satisfaction, and trust (Mohanty et al., 2020: 757). It plays an important role in raising the living standards of individuals with a combination of economic, technological, scientific, cultural, and organizational progress.

Realizing digital transformations within the sector will have an impact on business continuity. In destinations where tourism activities constitute an important source of income, digital transformations will provide support in ensuring income flow and increasing operational efficiency (UNWTO, 2020).

8.5 CONCLUSIONS

It is an undeniable fact that the epidemic process affects the tourism sector as well as every sector. In this respect, it is important to consider the short- and long-term changes that may occur in the sector. Although the epidemic period has negative reflections on the tourism sector, this crisis, which is experienced with the right timing and the right activities, can be turned into a good opportunity. However, it is necessary to adopt a sustainable tourism approach to prevent opportunities from turning into crises, just as crises turn into opportunities. As a matter of fact, one of the factors affecting the emergence of COVID-19 can be considered as the increase in imbalances in the ecosystem. The concept of sustainability is very important for maintaining this balance and for a more liveable world. Considering the intense population and energy use in the tourism sector, the importance of sustainability for tourism comes to the fore. If the tourism sector is considered in the triangle of sustainability and COVID-19, the new normal tourism policies should include measures to protect and improve the health of individuals, businesses, and society. At this point, the safe tourism certificate program pioneered by Turkey draws attention (TGA, 2021).

In addition to the socioeconomic effects of the COVID-19 epidemic, it can be said that the concept of sustainability has increased awareness especially on economic activities. Epidemic draws attention to the production and consumption of goods and services and the importance of local suppliers. In the cyclical economic process, recycling, waste management, etc., the monitoring of policies has an impact on economic activities. The implementation of activities such as supporting local food and preventing waste allows tourism businesses to increase their productivity (One Planet, 2021).

The main elements that stand out for sustainable tourism are explained as follows (Lanioglo and Rissanen, 2020: 523):

- Capacity
- Local Community
- Tourist Experience
- Destination Management and Organization
- Seasonality Management
- Demand Management
- Smart Destinations
- Resource Efficiency
- Physical Integration

The concept of sustainability, whose awareness has increased with the COVID-19 epidemic, is expected to come to the fore in the travel preferences of individuals. It is predicted that individuals will prefer more sustainable travels with an "eco-consciousness" mentality (Eco Bnb, 2021).

Post-COVID-19 tourism trends are predicted as follows:

- Increased interest in some types of tourism with a tendency to avoid crowded places: ecotourism, outdoor recreation, adventure tourism, etc.
- Less travel and longer stays,
- Increasing interest in different cultures,
- Traveling to closer areas with personal vehicles and accordingly development in the car rental sector,
- Increasing tendency to trust trains and private vehicles rather than airplanes,
- Increase in the trend of cycling,
- Increasing preference for small and boutique hotels,
- Increase in caravan, glamping, and camping preference,
- Increase in home tourism, villa rentals, yacht tours with small groups,
- Increase in digitalization trend,
- An increase in the tendency to winter holidays,

- An increase in the tendency to travel within its own region in international travels,
- Differences in health tourism preferences (tendency to more reliable countries with COVID 19 policies),
- Virtual tours (augmented reality, increase in virtual reality applications),
- The trend from popular tourism destinations to untouched and lesser known tourism destinations,
- Individuality in complementary touristic product preferences (individual tendencies in entertainment, orientation toward more individual activities such as diving, paragliding rather than disco and parties)
- Hygiene and sanitation practices are more decisive in providing competitive advantage among tourism enterprises,
- Decreased demand for the all-inclusive system (consumers tend to prefer personalized à la carte systems by decreasing their common dining preferences),
- The change in consumption and preferences in this direction due to the increase in ecological awareness in the consumer.

KEYWORDS

- **new normal policies**
- **tourism**
- **epidemic**

REFERENCES

Cakır, P; Barakazı, M. The Effect of the Coronavirus Process on the Tourism Sector and the Measures Taken Against the Epidemic. *Anadolu Univ. J. Soc. Sci.* **2020,** *20* (3), 313–332. DOI: 10.18037/ausbd.801802.

Calli, F. The Digital Future of the Tourism Industry. *J. New Tour. Trends (JOINNTT)* **2021,** *2* (1), 97–107.

Duzgun, E. New Tourist Choice After the Pandemic: Glamping Tourism. *Ordu Univ. Inst. Soc. Sci. J. Soc. Sci. Res.* **2021,** *11* (1), 145–158. DOI: 10.48146/odusobiad.870381.

Eco Bnb. https://ecobnb.com/blog/2021/04/impact-covid-tourism-travel-trends-tips/ (Accessed 5 July 2021).

Hall, C. M. On the Mobility of Tourism Mobilities. *Curr. Iss. Tour.* **2015,** *18* (1), 7–10.

Haque, T.H.; Haque, M.O. The Swine Flu and Its Impacts on Tourism in Brunei. *J. Hosp. Tour. Manage.* **2018,** *36*, 92–101.

Ianioglo, A.; Rissanen, M. Global Trends and Tourism Development in Peripheral Areas. *Scand. J. Hosp. Tour.* **2020,** *20* (5), 520–539. DOI: 10.1080/15022250.2020.1848620.

Ivanov, S.; Webster, C. Measuring the Impact of Tourism on Economic Growth. *Tour. Econ.* **2007,** *13* (3), 379–388.

Karabulut, G.; Bilgin, M.H.; Demir, E.; Doker, A. C. How Pandemics Affect Tourism: International Evidence. *Ann. Tour. Res.* **2020,** *84*, 1–5.

Kuo, H.; Chen, C.C.; Tseng, W.C.; Ju, L.F.; Huang, B.W. Assessing Impacts of SARS and Avian Flu on International Tourism Demand to Asia. *Tour. Manage.* **2008,** *29* (5), 917–928.

Lee, C. C.; Chen, C. J. The Reaction of Elderly Asian Tourists to Avian Influenza and SARS. *Tour. Manage.* **2011,** *32* (6), 1421–1422.

Maphanga, P. M.; Henama, U. S. The Tourism Impact of Ebola in Africa: Lessons on Crisis Management. *Afr. J. Hosp. Tour. Leisure* **2019,** *8* (3), 1–13.

McKercher, B.; Chon, K. The Over-Reaction to SARS and the Collapse of Asian Tourism. *Ann. Tour. Res.* **2004,** *31* (3), 716–719.

Mohanty, P.; Hassan, A.; Ekis, E. Augmented Reality for Relaunching Tourism Post-COVID-19: Socially Distant, Virtually Connected. *Worldwide Hosp. Tour. Themes* **2020,** *12* (6), 753–760. DOI: 10.1108/WHATT-07-2020-0073

OEF (Oxford Economic Forecasting). Will Swine Flu Push the World Into Deflation? *Econ. Outlook* **2009,** *33* (3), 13–17.

One Planet. https://www.oneplanetnetwork.org/sustainable-tourism/COVID-19-how-tourism-can-recover-responsibly (Access 22 June 2021).

Page, S.; Song, H.; Wu, D. C. Assessing the Impacts of the Global Economic Crisis and Swine Flu on Inbound Tourism Demand in the United Kingdom. *J. Travel Res.* **2012,** *51* (2), 142–153.

Skare, M.; Soriano, D. R.; Porada-Rochoń, M. Impact of COVID-19 on the Travel and Tourism Industry. *Technol. Forecast. Soc. Change* **2021,** *240*, 1–14. DOI: 10.1016/j.techfore.2020.120469.

TGA (Tourism Development Agency). https://www.tga.gov.tr/about-safe-tourism-program/ (Accessed 14 Nov 2021).

Topal, G. B. Digital Transformation in the Tourism Industry. In *Transformation of Professions and Industry in the Digital Future*; Öz, S., Onursal, F.S., CandanUca, N., Eds.; HiperlinkEğitim: İstanbul, 2020; pp 320–349. https://books.google.com.tr/books?hl=tr&lr=&id=iaACEAAAQBAJ&oi=fnd&pg=PT124&dq=cov%C4%B1d+19+Tourism+trends+2020&ots=0DsmKk9MZv&sig=nG2nQ3Do5oq05Fnj-fmK5xdFbK0&redir_esc=y#v=onepage&q&f=false (Accessed 01 July 2021).

TUYED (Association of Tourism Writers and Journalist). http://www.tuyed.org.tr/serif-yenenden-2021-turizm-trendleri/ (Accessed 01 July 2021).

TURSAB (Association of Turkish Travel Agency). https://www.tursab.org.tr/edergi?pdf=/assets/assets/uploads/arastirmalar/koronavirus-r-mayis-2020.pdf (Accessed 01 July 2021).

TURSAB (Association of Turkish Travel Agency). https://www.tursab.org.tr/edergi?pdf=/assets/assets/uploads/arastirmalar/COVID-19-surecinde-turkiye-ve-dunya-turizmi.pdf (Accessed 01 July 2021).

UNWTO (World Tourism Organization). *UNWTO BriefingNote—Tourismand COVID-19, Issue 1—How Are Countries Supporting Tourism Recovery?* UNWTO, Madrid. https://doi.org/10.18111/9789284421893 (Accessed 21 June 2021).
UNWTO (World Tourism Organization). https://www.unwto.org/COVID-19-oneplanet-responsible-recovery. (Accessed 22 June 2021).
UNWTO (World Tourism Organization). https://www.unwto.org/investments; https://webunwto.s3.eu-west-1.amazonaws.com/s3fs-public/2020-09/FDI-Tourism-Report-2020.pdf (Accessed 22 June 2021).
UNWTO (World Tourism Organization). https://www.unwto.org/news/un-policy-brief-on-tourism-and-COVID-19 (Accessed 22 June 2021).
Unluonen, K.; Ceti, B. Evaluation of the Effects of the Crises Caused by Epidemics on the Tourism Sector. *Ankara HacıBayramVeli Univ. J. Tour. Faculty* **2019,** *22* (2), 109–128.
Whaley, F.; Mansoor, O. D. SARS Chronology. In *SARS: How a Global Epidemic Was Stopped*; World Health Organization, Ed.; WHO Library, 2006; pp 3–47.
WHO. (World Health Organization). *Avian Influenza Weekly Update Number 818.* https://www.who.int/docs/default-source/wpro---documents/emergency/surveillance/avian-influenza/ai-202111112.pdf (Accessed 14 Nov 2021).
WHO. (World Health Organization). *Ebola Virus Disease.* https://www.who.int/news-room/fact-sheets/detail/ebola-virus-disease (Accessed 14 Nov 2021).

CHAPTER 9

Analyzing Microfinance as a Resilience Strategy for Poverty Alleviation and Promoting Sustainable Tourism: A Communicative Perspective

MANPREET ARORA[1] and ROSHAN LAL SHARMA[2]

[1]*HPKVBS, School of Commerce and Management Studies, Central University of Himachal Pradesh, Dharamshala, Himachal Pradesh, India*

[2]*Department of English, Central University of Himachal Pradesh, Dharamshala, Himachal Pradesh, India*

ABSTRACT

Microfinance as a tool for poverty reduction and sustainable development of tourism has received extensive recognition in various parts of the world. Microfinancing activities have shown a tremendous impact on various developing countries like Bangladesh, India, Bolivia, Cambodia, Mangolia, Indonesia, and Philippines. The microfinance programs have proved to be a successful tool in reaching out to the poor and solving issues like poverty, removing gender disparity, unemployment, regional imbalances, alongside boosting tourism industry across the world. In India also, microfinance has proved to be effective and successful in transforming lives of the poor people residing in rural areas without access to financial services and in the process enhancing tourism potential through encouraging nature-based tourist activities, particularly in the regions having natural grandeur. If these programs are aligned with effective management strategies and if more emphasis is laid

Dynamics of the Tourism Industry: Post-Pandemic and Post-Disaster Perspectives and Strategies.
Gül Erkol Bayram and Anukrati Sharma (Eds.)

on their communicative dimensions, these can promote, as well as enhance, tourism and entrepreneurial activities in the socially excluded sectors of the society, and thereby connect them with rest of the world. Moreover, if certain NGOs come forward in effective training and development of the neglected parts of the society, it can work wonders. Communication with the marginalized/excluded sections of societies has been lacking at various levels in India. It is also true about people from various communities, castes, tribes, and societies across the globe. To connect with the marginalized with a view to benefit them, effective communicative strategies can surely play a great role in creating awareness about microfinance/microcredit. This chapter is a qualitative piece of work and its theoretical foundation is based on the review of literature and secondary data to support and develop a conceptual perspective. Moreover, an attempt has also been made to create a theoretical model to guide the policy makers or voluntary organizations in coming forward to initiate/support activities concerning empowerment of poor/deprived chunk of rural population through microfinance.

9.1 INTRODUCTION

The developing nations across the world have been facing challenges concerning poverty, unemployment, hunger, and of removing inequality and injustice. After facing several waves of the COVID-19 pandemic, the developing countries need to focus on resilience strategies in every sector to make their economies functional in every arena. Beyond the traditional ways of managing the economies, mixed models in every sector are emerging to gain advantage of late. Considering the efforts of Nobel Laureate Muhammad Yunus, who established the Grameen Bank in Bangladesh and developed the concept of microfinance and microcredit for the betterment of the improvised/marginalized sections of the society and particularly women, a multi-sectoral development approach attached to microfinance can have tremendous possibilities of change. He was of the view that granting loans to vulnerable, poorer, and under-privileged sections of the society, along with women, can result in better rate of return as it helps support their families and community, and eventually lead to poverty alleviation and female empowerment. Alleviation of poverty and women empowerment are not the only objectives; the basic thrust instead is on making them self-sufficient and reliable breadearners for their families. His work and the concept of microfinance which focuses on the upliftment of the marginalized sections of the society was imitated in various countries and yielded positive results. The National Bank for Agriculture and Rural Development (NABARD) has played an important role in India in successfully establishing the microfinance program with the help of various stakeholders especially the nongovernmental organizations, banks, and cooperatives in formal and informal sectors. NABARD launched the Self-help Group (SHG) Bank Linkage Programme as its pilot project to encourage the microcredit activities across rural grassroots level of India. It was focused on adequately addressing the microcredit needs of the poor, and especially of the women. The positive results reaped by this pilot project gave motivation to the government and the scheme was further extended in every nook and corner of India as it became a national priority. Various union budgets focused on providing microcredit in India. It became an elementary model for uplifting the rural society where the formal financial services were not available to the poor.

Microfinance institutions, along with various nonbanking institutions, trusts, and cooperative banks, have employed a variety of measures to ensure the delivery of microcredit, especially at the ground level in rural areas. Initially, the Small Industries Development Bank of India (SIDBI) and NABARD were the main players in providing microcredit along with certain cooperative banks. In the later stages, however, the commercial banks were linked with Self-help Groups (SHGs) to avail credit especially at the rural levels. The Bank Linkage Programme has played an instrumental role in not only the policy formulation of provision of microcredit but it has also undertaken various kinds of interventions at financial levels. Worldwide microfinance has been regarded as an important tool to achieve better growth rate (Bangoura, 2012) and bring the marginalized sections of the society to the mainstream. It has also proven to be a means of poverty reduction (Razzaque, 2010) and encouraging entrepreneurial activities at micro, small, and medium-scale levels. Poor people who do not have access to finance are largely benefited due to microfinance programs (Milgram, 2001). Green microfinance is yet another sector which is gaining prominence in the field of encouraging sustainability (Forcella and Hudon, 2016). In this chapter, we argue that it can prove to be a wonderful tool to include the socially excluded/marginalized sections of the society at various levels in the mainstream, and thus empower them. Without confining to this view, we also believe that if microfinance is provided to the local people especially residing in the tourist destinations, it can promote sustainable tourism; generate income for localities; and encourage local art, culture, food products, and services as ethnic and indigenous things always have the potential to attract tourists.

This, in turn, can boost both tourism sector as well as rural entrepreneurship besides having a ripple effect in the form of promoting local art, skills, revival of culturally grounded traditional products, and services. Finance is the major hurdle in every venture or entrepreneurial activity. To boost any kind of service, financial resources are required. Therefore, the government in collaboration with the commercial banks, financial institutions, development banks, funding agencies, and NGOs can focus mainly on microcredit interventions in tourism sector. That is how the possibilities of revival of this sector can be increased. The ultimate goal of achieving the sustainable development goals (SDGs) can be linked to developing a plan of action in various sectors via coalescing them with microcredit interventions.

9.2 LITERATURE REVIEW

It can be arguably proven that in case the formal financial services are not able to reach the poorest of the poor, they remain excluded from the mainstream. Like several developing countries, in India too, there are various vulnerable sections in the society that remain socially excluded (Singh, 2014). In fact, we have a long history to evidence that many volunteers, social workers, and political leaders have fought for various socially secluded/isolated sections of the society. One of these deprived sections is the women (Srivastava and Srivastava, 2010; Menon, 2009; Chakrabarty, 2012). Rural tribal women still stand marginalized in states like Tripura (Ghosh and Choudhuri, 2011) as they are socially excluded. But they have indigenous knowledge, art, and rich traditions to which tourists get attracted. When the society is not able to offer everything to all the sections, groups, and individuals, they are said to be socially excluded, and women are one of them (Power, 2000). Evidences are also available which show that certain groups of the society remain poor and the proportion is not equitable as they remain uneducated and deprived due to discriminations in the society (Thorat and Attewell, 2007).

Being deprived of something is in many ways related to social exclusion. Both have different dimensions and are concerned with groups as well as individuals (Bossert et al., 2007). There can be many bases of exclusion like "age, caste, gender, disability, ethnic background, religion, etc" (Borooah, 2010). Evidences in countries like the European Union show that in case individuals are not economically empowered or are deprived and socially excluded, they derive less satisfaction from life (Bellani and D' Ambrosio, 2011). Those who are materially deprived, their "physical health" and "social relationships" are also affected adversely and deprivation also affects their quality of life (Bayram et al., 2012). Various developing nations like India, Bangladesh, and Turkey show that when deprived sections of the society are socially excluded, it affects their self-esteem negatively (Khan et al., 2009). Social exclusion can even lead to higher cases of suicides. Evidences in countries like Japan are available where social exclusion is one of the causes of high representation of males above the age of 50 who commit suicide and are homeless (ABE, 2010).

The concept of empowerment has a wide scope and it has been defined and described by various authors differently. In a general sense, however, a woman is said to be empowered when she is able to control various decision-making activities in her professional and personal life. They get socially uplifted too as it increases their motivation level, helps them become self-dependent, and increases their ability to take decisions rationally at the family level, like decisions related to education of children, spending of money, purchase of assets, travel, etc. They become more aware of socio-cultural issues around them and participate in various activities and ceremonies. Evidence of increase in bargaining power of women is also available due to microfinance. Many women become politically aware also and achieve greater autonomy at the societal level. Studies emphasizing social impact of microfinance also highlight that women could raise their voices against social evils like child marriage, female foeticide, drug abuse, and domestic violence. In India particularly, the Self-help Group Bank Linkage Programme has helped in alleviating poverty and empowering women to join the mainstream (Arora and Singh, 2016). The poor and marginalized sections of the society have been empowered at various levels when they are given access to credit. It helps them not only to uplift their status in the society, but also paves way for their autonomy in decision-making at various levels in their personal as well as professional lives. Financial services are not just limited to savings and disbursement of credit. These are also linked to sectors like insurance, investments, and asset holdings (Errais and Miled, 2015). The SHG Bank Linkage Programme has acted as a bridge to fill the gap by providing them opportunities to formally get linked to all such financial services which in turn help them attain empowerment in several ways. The same can be true if this phenomenon is extensively used in tourism sector which has been badly hit due to the COVID-19 pandemic. The in-house tourist activity can promote the revival of this sector especially in nature-based tourism, sustainable tourism, and rural tourism. Sustainable tourism has been regarded as an effective tool for eliminating poverty (Zhao and Ritchie, 2007). Promoting inbound tourism can also be an effective resilience strategy as by promoting local fairs and festivals, the people who can be engaged in different skill-based activities can benefit considerably (Mensah and Boakye, 2021).

9.3 RESEARCH METHODOLOGY AND RESEARCH QUESTIONS

More than 70% of the world's poor women have been underprivileged, disadvantaged, and vulnerable and they never had an access to microcredit and other services provided by financial institutions. In every country, commercial banks are formal, business, and profit-making institutions. Their main focus is on men and women who can provide them collateral securities and help them earn profits. But at the poorest of the poor level, they do not have many types of collateral with them to keep with the commercial banks in order to get credit. This is true for tourism service providers too. Therefore, microfinance is regarded as a special intervention program which does not require any collateral at any stage and has proven to be an effective tool to uplift the masses economically and socially. This chapter seeks to get answers to the following research questions qualitatively by using the secondary data and literature support:

1. How far have the poor people been benefitted by the microfinance programs, and to what extent have socially excluded women been included in the mainstream economically?
2. What is the extent of disbursement of credit to women in India and what is the position of microcredit disbursement to the poor people in various parts of India?
3. How can coalescing tourism industry and microfinance be fruitful for women and vulnerable sections of the society?
4. How can effective communication help in uplifting women rationally?

9.4 STATUS OF MICROFINANCE IN INDIA

Before pondering over the sector-specific interventions, needs, and strategies, let us explore the status of microfinance in India as a whole. This data focuses on total microfinance lending and disbursements irrespective of any sector. In India, the microfinance industry has catered to the credit needs of the borrowers to a great extent. Till June 30, 2020, 10.37 crore of loan accounts were served (bfsi.eletsonline.com). Undoubtedly, the COVID-19 related lockdown in the country affected credit payback, but microfinance borrowers were offered a period of moratorium as well. Special liquidity facility was provided to microfinance institutions to deal with the pressures of the lockdown during the pandemic. The provision of microcredit can prove to be a wonderful coping strategy as part of resilience strategy in the economy. The only solution to bring back the economy on track and deal with the issues of poverty, hunger, and unemployment can be the encouragement to self-help groups (SHGs) and small startups and aid to micro, small, and medium-scale enterprises. The concept of self-help group has been very successful in various states of India. The recovery rate became an issue due to the pandemic but as the spending of people got an upsurge, the recovery rates also improved in India. An improvement in collection efficiency as well as disbursements were seen to be steadily growing (Roy, 2022; HINDUBUSINESSLINE).

The microfinance provision in the form of microcredit plays a very important role in promoting/encouraging the inclusive growth of the vulnerable sections of the society that are marginalized and excluded, especially the poorest of the poor and women. The borrowers are mostly the ones who are at the bottom of the pyramid, and for them availing credit directly from commercial banks is a costly affair. They are practically excluded from the mainstream as they have no collateral to offer for the profit-earning institutions to avail credit. Evidence can be found where microfinance sector has been instrumental in generating varied business opportunities for below-poverty level and very low-income households (Khandker and Samad, 2014; Pantelić, 2011; Kleynjans and Hudon, 2016). These opportunities are generally out of the reach of traditional financial services offered by various categories of banks and other financial institutions. In India, microfinance sector provided a great boost to the unorganized sector. It has proved to be a successful tool for generation of self-employment and creation of jobs in various sectors. The tourism sector has the potential of improving the microscale small tourism-based enterprises and the economic effects of such pro-tourism enterprises are many including eradication of poverty (Zhao, 2009).

Eradication of poverty is the biggest challenge that developing countries face across the world. Literature through empirical evidence suggests that microfinance is an effective policy instrument to help eradicate poverty (Satish, 2005; Tavanti, 2013; Deedam and Onoja, 2015; Vasimalai and Narender, 2007). One of the factors which contribute toward poverty is limited or no access to credit facility (Mohamed and Temu, 2008). Therefore, in several countries, provision of sustainable microfinance helps in reduction of poverty (Arora and Singh, 2022). A sizeable chunk of literature on microfinance suggests that if access of finance is provided to the vulnerable sections of the society, it can help reduce poverty substantially (Galab and Chandrasekhara, 2003; Haq, 2015; Berhane and Gardebroek, 2011). When poor sections of the society receive access to finance, it can contribute via earning income for a long period of time as the amount is invested in income-generating activities which help increase and diversify the sources of income. It also helps in the increase of the borrowers' assets and further increases level of consumption by improving their standard of living (Khavul, 2010; Sinha, 2005; Basu and Srivastava, 2005; Schrieder and Sharma, 1999; Bhuiya et al., 2016). It is also helpful in reducing the risk of borrowers associated with illness and natural calamities and thus contributes toward their health and wellbeing along with that of their families. Economic and social empowerment of the borrowers substantially improves due to the access of credit (Qudrat-I Elahi, 2003; Khandelwal, 2007; Sharma, 2011; Vatta, 2003; Karim, 2014; Alldén, 2009). Women have always contributed toward economic growth and sustainable livelihood of their families and communities across the world. In microfinance programs as well, women have comparatively benefited more than men (Chhay, 2011; Leach et. al 2002; Hartarska et al., 2014; Pitt et al., 2006; Sooryamoorthy, 2005; Jafree and Ahmad, 2013; Scheyvens, 2015; Vonderlack and Schreiner, 2002).

Women have been the main focus while disbursing credit under the SHG Bank Linkage Programme (SHGBLP) in India (Arora and Singh, 2016). Since 1992, the amount of credit disbursed has shown a favorable trend toward women in almost all the states of India. Savings and investment are the key drivers of economic growth in every economy. The amount of savings in the banks due to microfinance activities shows a positive increase during the last decade. The number of all women SHGs has continuously increased from 2015 to 2018. Table 9.1 shows that out of total SHGs which save in bank, the percentage of women SHGs demonstrate a considerable increase, which simply implies that more credit to women paves their way toward inclusion and development, and that it helps substantially in poverty alleviation.

As evident, the percentage of women SHGs out of total SHGs was 85.58% in 2015–2016 but in 2016–2017 it was 85.63% compared to 84.51% in 2017–2018. The loans disbursed to all women SHGs also show a favorable trend toward women in these years (Arora and Singh, 2018). This is owing

TABLE 9.1 SHG Savings with Banks and Loans Disbursed to SHGs in India.

Particulars		2015–2016		2016–2017		2017–2018	
		No. of SHGs	**Amount**	**No. of SHGs**	**Amount**	**No. of SHGs**	**Amount**
SHG savings with banks as on March 31	Total SHG Nos.	79.03 (2.68%)	13691.39 (23.79%)	85.77 (8.53%)	16114.23 (17.69%)	87.44 (1.95%)	19592.12 (21.59%)
	All women SHGs	67.63 (1.68%)	12035.78 (29.92%)	73.22 (8.26%)	14283.42 (18.67%)	73.90 (0.94%)	17497.86 (22.51%)
	Percentage of women	85.58	87.91	85.36	88.64	84.51	89.31
	Of which NRLM/SGSY	34.57 (13.27%)	6244.97 (41.16%)	37.44 (8.30%)	7552.70 (20.94%)	41.84 (11.76%)	10434.03 (38.15%
	% of NRLM/SGSY groups	43.74	45.61	43.65	46.87	47.85	53.26
	Of which NULM/SJSRY	4.46 (3.00%)	1006.22 (6.12%)	5.46 (22.42%	1126.86 (11.99%)	4.25 (−22.10%)	1350.80 (19.87%)
	% of NULM/SJSRY groups	5.64	7.35	6.36	6.99	4.86	6.89
Loans Disbursed to SHGs during the year	Total no. of SHGs extended loans	18.32 (12.67%)	37286.90 (35.18%)	18.98 (3.60%)	38781.16 (4.01%)	22.61 (19.13%)	47185.88 (21.67%)
	All women SHGs	16.29 (12.50%)	34411.42 (40.92%)	17.16 (5.34%)	36103.13 (4.92%	20.75 (20.92%)	44558.74 (23.42%)
	Percentage of women groups	88.92	92.29	90.42	93.09	91.78	94.43

Source: Adapted from Annual Reports on Microfinance (NABARD.COM, 2021).

to the fact that empowerment of women is the empowerment of the whole community as its effect percolates to the lowest levels where the poorest of the poor benefit concretely. NABARD has also taken initiatives for tribal women by providing them facilities of SHGBLP. A NABARD-supported "Rural Mart" has been established for linking tribal women at Chamba in Himachal Pradesh under the brand name of "Pangihills at Chamba, Himachal Pradesh." The mart is helping women of self-help groups, farmers, artisans, etc., in value chain management activities as they suffer from absence of infrastructure facilities. They face a lot of challenges in marketing of their products related to "agricultural produce, horticulture and forest-based produce." The impact has been establishing a sustainable livelihood for tribal women of Chamba (NABARD, 2019–2020). The region-wise progress of savings-linked SHGs with banks had been also quite impressive which focuses on balanced regional development through microfinancing activities. The region-wise progress can be shown with the help of Table 9.2 from the year 2017–2018 to 2019–2020.

TABLE 9.2 Region-wise Progress of Saving-linked SHGS with Banks (2017–2018 to 2019–2020).

S. No	Regions	2017–2018		2018–2019		2019–2020	
		No. of SHGs	Savings-amount	No. of SHGs	Savings-amount	No. of SHGs	Savings-amount
1	Northern region	478883	49293.91	548624	62452.82	577122	59549.73
2	North Eastern region	485591	32207.59	523469	40407.05	556899	48140.55
3	Eastern region	2130997	441803.18	2654358	601154.88	2811130	664332.73
4	Central region	902222	95385.11	1062759	133230.00	1135083	171217.00
5	Western region	1097448	124694.93	1388615	205275.15	1473853	201880.14
6	Southern region	3649296	1215826.80	3836418	1289928.25	3689236	1470084.74
	Total	8744437	1959211.52	10014243	2332448.15	10243323	2615204.89

Source: Adapted from Annual Reports on Microfinance (NABARD.COM, 2021).

The highest share of number of SHGs formed is of the southern region. The number of SHGs in the southern region was 3649296 in 2017–2018 compared to 3836418 in 2018–2019, which showed a slight decrease in the year 2019–2020 and was 3689236. But, overall, the contribution of the southern region has always been highest followed by eastern and western regions.

9.5 STRATEGIZING MICROFINANCE AS A TOOL TO ALLEVIATE POVERTY AND ENCOURAGING TOURISM

The importance of microfinance in almost every industry, including tourism, is significant. In fact, the role of microfinance has proven to be of utmost importance in boosting the small-scale sector. It helps dispense the access of the capital to budding entrepreneurs, which ultimately becomes part of the small and medium-size industry. Combined with tourism, it gives tremendous opportunities to allied goods and service sector which go parallel with tourist activities. The ultimate aim of the microfinance in this sector is to help the tourists get opportunity to know the real culture, traditions, and customs of the destination through local goods, services, souvenirs, foods/cuisines, and markets. The easy credit facilities to these ventures, which also help in sustainable tourism by conserving local traditions and customs, help the local people earn their livelihood by selling local products to tourists. The products that are generally available locally involve local raw material and human interventions which are good for the economy as a whole as people become self-sufficient and job-providers rather than job-seekers. They also help in boosting rural and sustainable tourism. Another facet of such local markets can be seen at various pilgrimage destinations where souvenirs, metal products, idols, and related goods are sold in the market which boost self-sufficiency and earning opportunities for the locals who are skilled or semiskilled artisans. Microfinancing all such ventures/activities has a strong influence in promoting sustainability via encouraging local traditions and culture.

Microfinance and microcredit are arguably the most potent means to encourage sustainable entrepreneurial ventures that are sensitive toward nature, ecology, and environment and that has the potential to encourage and increase tourists' footfall. In a country like India, microcredit from various banks and microfinance institutions has gained considerable momentum during the recent decades. Microfinance and tourism share a symbiotic relationship characterized by interesting give-and-take which results in financial empowerment of people from low-income groups and from areas less developed from the viewpoint of tourism and tourist activity. The capital involved in such ventures is not huge; on the contrary, it is miniscule but the dividends that it offers have immense significance.

9.6 THE COMMUNICATIVE DIMENSION

It is commonly known as to how the poorest of the poor are marginalized everywhere including in the rural areas. Women, in fact, are doubly marginalized as firstly they are poor/deprived and hail from the marginalized section of the society, and secondly because they are further exploited and ill-treated within that marginalized space by their male counterparts due to the prevailing patriarchal mindset. Just because they happen to be female of the species, women have to face double jeopardy and marginalization at several levels. Economic disempowerment too is, in several ways, an offshoot of the injustice meted out to them not just by menfolk but also by their own ilk (i.e., women) as they too play the second fiddle to the men in their families/community. More often than not, we find women least self-reliant financially in the villages as there are inherent biases/prejudices vis-à-vis their ability, potential, and capability as the ones who could manage family and financial affairs. The fact of the matter, however, is that despite their financial disempowerment, women serve as the backbone of the households that they manage. By and large, the menfolk in rural areas tend to control whatever little money they have with the result that women, despite being the actual contributors to the family income (howsoever small), remain deprived of both owning as well as managing family income from agricultural produce or other sources. And that precisely is where microfinance and microcredit as facilitating and enabling mechanisms in rural areas assume critical importance.

Being a service provided by banks to help impoverished/ignored/marginalized communities, microfinance has an important communicative dimension, which is germane to the issue of poverty alleviation and women empowerment in rural areas. To create awareness among people about the meaning, nature, and significance of microfinance, and to communicate the same to poor rural folks in general and women in particular remains an uphill task. The lending institutions find it hard to convince people to voluntarily go for microcredit/small loans. The naïve rural mindset is steeped in beliefs such as the richest person is the one who does not owe a penny to anyone, and this is what gets etched in their psyche. They do understand what mutual co-operation is all about but usually remain free from matters concerning lending/borrowing money. Microfinance as a way of empowering people to become financially self-reliant, by and large, eludes the understanding of the poor rural folks. To make them understand as to how microfinance can help them attain self-reliance and overcome their penury-struck and impoverished condition becomes really challenging. Microfinance-related practices therefore have really gained ground in developing countries in a big way as data show that millions of people have come out of their poverty due to such practices.

Microfinance as a communicative strategy can be viewed as language of empowerment as well. This is because of the fact that micro-financing the socio-culturally marginalized and economically deprived sections of the society leads to their financial self-dependence and freedom. The amount of money which is disbursed to the poor people is fairly small but the impact that it has in the form of poverty alleviation is enormous. One needs to view it critically and at a greater depth. To consider microfinance as language cannot in any sense be erroneous. "Microfinance" as a term is a combination of two words namely "micro" and "finance" implying "small" and " money," respectively. The first challenge that the lending institutions such as banks or cooperatives face in making people understand the basic implication of the term "microfinance" has got a lot to do with the basic nonreceptive mindset of poor people in rural areas who are steeped in a particular kind of belief system. This, as stated earlier, is because of the reason that village folks do not believe in borrowing money. Moreover, they consider it even unethical to remain in debt. So much so, indebted identity of person is equated with unethical code of moral conduct in life. Interestingly, in the villages, the richest person is the one who does not owe money to anyone, or does not remain under the psychological burden of repayment of the money borrowed from a person or a bank. Amid this kind of prevailing mindset, lending institutions such as rural banks, cooperative banks, commercial banks, and other agencies face the herculean task of making the rural and poor population aware of the importance of microfinance and its benefits, and this is where communication becomes extraordinarily critical.

The communication that a financial institution has to resort to while communicating with prospective customers in the form of borrowers from the villages has got to be effective. Simple communication cannot be very successful/effective in such a scenario. On the contrary, lending institutions have to devise special communicative strategies to: (a) decondition the conventional rural mindset regarding borrowing money or accepting financial assistance; (b) motivate/persuade the rural folks to come out of conventional grooves of habit regarding accepting loans from financial institutions ; (c) make them understand how important their role is as borrowers/customers of the bank, and also make them realize their importance as major stakeholders in all the schemes relating to micro-financing practices; (d) create awareness about their changed behavior as borrowers/customers particularly from the point of view of their continual engagement or involvement with the bank; and (e) give them to understand that as borrowers/customers are the backbone of the whole microfinance/microcredit system, and thereby also make them aware of their rights as well as responsibilities as the key stakeholders in the entire process.

To achieve the stated objectives, the lending institutions need to take recourse to appropriate communicative strategies that are effective, persuasive, convincing, successful, and sustainable. Lending institutions, it has been observed, have been coming forward in a big way to lend microfinancial help to women SHGs. This has proven to be a great strategy to empower the disempowered segments of female population. The microfinance cuts across gender categories and thereby help men as well as women, but when we think in terms of SHGs, we aim at addressing the problems of female population in particular. In both the cases, communication serves as the locus. One of the objectives that microfinance institutions must achieve is educating the rural entrepreneurs to hone and utilize the tourism potential in their respective locations. This is because of the fact that tourism industry has been making inroads into diverse geographical spaces; however, rural spaces occupy the place of prominence. The microfinance institutions must communicate to the poorest of the poor the benefits of nature-based and countryside tourism and the importance of locally produced goods and cuisines. Once the rural population understands its rich financial implications, they may take lead in rural entrepreneurial ventures that may help them overcome their financial hardships and empower them in a way that they become assets via contributing incessantly and sustainably to tourism industry in times to come.

Thus, as discussed earlier, the basis of entire micro-financial activity is communication. While strategizing such communication, the lending/microfinancing institutions need to keep in mind the following:

1. The language of communication should be lucid and clear.
2. The communication should be factually correct. It should be persuasive and not coercive.
3. It has to be persistent (on continual basis) and not occasional.
4. Such a communication should aim at achieving the objectives of lending institutions and thereby benefit the rural borrowers to maximum possible extent.
5. The communication should be effective in the sense that the message should successfully get across to rural borrowers/customers.

The beneficiaries of microfinance need to be constantly mentored and counselled. Moreover, they require training as well regarding how they have to conduct themselves as the responsible customers/stakeholders in the microfinance system. In order to train the borrowers, financial institutions need to come up to their expectations. For instance, if the representatives of financial institutions are not able to go down to the level of the borrowers, the latter will not be able to understand the former's viewpoint/vision. Therefore, it is essential to make sure that the communication gap between the representative of the lending institution and the borrowers is minimized without any lapse. Once you have won the confidence of the borrower and once your creditability is established as a representative of lending institution, the microfinance-related communication becomes a cake walk. The point is not that of just preparing the borrowers to borrow the money; on the contrary, it is to make them feel empowered and responsible at the same time vis-à-vis how the money is to be utilized, in what manner, and with what end in view. In other words, the objectives of microfinance have to be clear on the part of both the lender as well as the borrower.

9.7 CONCLUSION

To sum up, we can say that microfinance is arguably the most potent tool to alleviate poverty, enable rural entrepreneurs to start nature-based tourist activities, and empower the marginalized sections of the rural population who have to face social exclusion particularly because they are not acknowledged as the major stakeholders in generating income in the rural households. To generate family income and to manage it in such a way that the bane of poverty is obliterated, microfinancing institutions have to make them significant stakeholders in decision-making at the socio-cultural as well as economic levels. The diversified utilization of microcredit may serve as the key to lend agency to the poor who are forced to remain at the lowest rung of the socio-economic ladder. Empowering the rural folks financially in the form of lending small loans to them and then helping them utilize the loan amount in initiating small entrepreneurial activities such as the ones concerning rural and nature-based tourism as mentioned earlier, will give the women direct control of the capital/money which they feel is theirs and which may further nurture as well as boost their entrepreneurial spirit. There is ample data to show as to how microfinance has empowered the poor folks in rural areas by enabling them to overcome poverty and economic disparity. It fills them with self-esteem to realize the importance of their potential and capability.

Equally crucial is the implementation of microfinance programs, which cannot be ensured if we lack in our understanding of the communicative dimension of such services provided by various funding agencies/microfinancing institutions. Reaching out to the borrower who hails from the poorest sections of the society is the most important factor in the entire marketing process of such programs, and hence, it remains a constant worry with lending agencies/institutions. That is why communication with the marginalized/excluded sections of the society becomes critically crucial. Communication serves as the basis to make any microfinance-related activity successful in rural areas. Creating awareness among the rural male/female entrepreneurs as to how they can come forward to accept small loans through microfinance responsibly can be best done through employing communicative strategies that, rather than intimidating the poor/disempowered borrowers, convince them via instilling confidence in them (through one-to-one meetings or small target-oriented workshops) and thus winning their trust. Once this is done, they can participate enthusiastically in microfinance-related activities and thus improve their economic condition to eventually emerge as empowered and responsible stakeholders in their family, community, and the society at large.

KEYWORDS

- **microfinance**
- **poverty**
- **social exclusion**
- **social inclusion**
- **resilience strategy**
- **microcredit**
- **communication**
- **communicative strategies**

REFERENCES

ABE, A. K. Social Exclusion and Earlier Disadvantages: An Empirical Study of Poverty and Social Exclusion in Japan. *Soc. Sci. Jpn J.* **2019,** *13* (1), 5–30. http://www.jstor.org/stable/40961239

ALLDÉN, S. Microfinance and Post-conflict Development in Cambodia and Timor-Leste. *Sojourn: J. Soc. Iss. Southeast Asia* **2009,** *24* (2), 269–284. http://www.jstor.org/stable/41308327

Arora, M.; Singh, S. Disbursement of Credit Under the SGSY Scheme: A Comparison of SHGs'swarozgaris and Individual Swarozgaris. *Indian J. Finance* **2016,** *10* (11), 54–63.

Arora, M.; Singh, S. Microfinance for Achieving Sustainable Development Goals: Pondering Over Indian Experiences for the Preservation of Magnificent African Natural Resources. In *Microfinance and Sustainable Development in Africa*; IGI Global, 2022; pp 82–102.

Arora, M.; Singh, S. Microfinance, Women Empowerment, and Transformational Leadership: A Study of Himachal Pradesh. *Int. J. Leadership* **2018,** *6* (2), 23.

Bangoura, L. Microfinance as an Approach to Development in Low Income Countries. *Bangladesh Dev. Stud.* **2012,** *35* (4), 87–111. http://www.jstor.org/stable/41968844

Basu, P.; Srivastava, P. Exploring Possibilities: Microfinance and Rural Credit Access for the Poor in India. *Econ. Polit. Wkly* **2005,** *40* (17), 1747–1756. http://www.jstor.org/stable/4416534

Bayram, N.; Bilgel, F.; Bilgel, N. G. Social Exclusion and Quality of Life: An Empirical Study from Turkey. *Soc. Indicat. Res.* **2012,** *105* (1), 109–120. http://www.jstor.org/stable/41409404

Bellani, L.; D' Ambrosio, C. Deprivation, Social Exclusion and Subjective Well-Being. *Soc. Indicat. Res.* **2011,** *104* (1), 67–86. http://www.jstor.org/stable/41476541.

Berhane, G.; Gardebroek, C. Does Microfinance Reduce Rural Poverty? Evidence Based on Household Panel Data from Northern Ethiopia. *Am. J. Agric. Econ.* **2011,** *93* (1), 43–55. http://www.jstor.org/stable/41240259

Bhuiya, M. M. M.; Khanam, R.; Rahman, M. M..; Nghiem, H. S. Impact of Microfinance on Household Income and Consumption in Bangladesh: Empirical Evidence from a Quasi-Experimental Survey. *J. Dev. Areas* **2016,** *50* (3), 305–318. http://www.jstor.org/stable/24737432

Borooah, V. K. Social Exclusion and Jobs Reservation in India. *Econ. Polit. Wkly.* **2010,** *45* (52), 31–35. http://www.jstor.org/stable/27917960.

Bossert, W.; Conchita D.; Vito P. Deprivation and Social Exclusion. *Economica* **2007,** *74* (296), 777–803. http://www.jstor.org/stable/4541571

Chakrabarty, M. Women Empowerment in Assam: A Study. *Indian J. Polit. Sci.* **2012,** *73* (1), 97–100. http://www.jstor.org/stable/41856564

Chhay, D. Women's Economic Empowerment Through Microfinance in Cambodia. *Dev. Practice* **2011,** *21* (8), 1122–1137. http://www.jstor.org/stable/41413008

Deedam, N. J.; Onoja, A. O. Impact of Poverty Alleviation Programmes on Indigenous Women's Economic Empowerment in Nigeria: Evidence From Port Harcourt Metropolis. *Consilience* **2015,** *14*, 90–105. http://www.jstor.org/stable/26188743

Errais, E.; Miled, S. B. Investing in Microfinance: The Case of Tunisia. *J. Private Equity* **2015,** *18* (3), 79–96. http://www.jstor.org/stable/43503855

Forcella, D.; Hudon, M. (2016). Green Microfinance in Europe. *J. Busi. Ethics* **2016,** *135* (3), 445–459. http://www.jstor.org/stable/24736065

Frances, S. Access, Use and Contribution of Microfinance in India: Findings from a National Study. *Econ. Polit. Wkly*. **2005,** *40* (17), 1714–1719. http://www.jstor.org/stable/4416529

Galab, S.; Chandrasekhara R. N. Women's Self-Help Groups, Poverty Alleviation and Empowerment. *Econ. Polit. Wkly* **2003,** *38* (12/13), 1274–1283. http://www.jstor.org/stable/4413378

Ghosh, B.; Choudhuri, T. Gender, Space and Development: Tribal Women in Tripura. *Econ. Polit. Wkly.* **2011,** *46* (16), 74–78. http://www.jstor.org/stable/41152108

Haq, R. Quantifying Vulnerability to Poverty in a Developing Economy. *Pak. Dev. Rev.* **2015,** *54* (4), 915–929. http://www.jstor.org/stable/43831374

Hartarska, V.; Nadolnyak, D.; Mersland, R. Are Women Better Bankers to the Poor? Evidence from Rural Microfinance Institutions. *Am. J. Agric. Econ.* **2014,** *96* (5), 1291–1306. http://www.jstor.org/stable/24476976

Jafree, S. R.; Ahmad, K. Women Borrowers of Microfinance: An Urban Lahore Study. *J. Third World Stud.* **2013,** *30* (2), 151–172. http://www.jstor.org/stable/45198686

Kamal Vatta. Microfinance and Poverty Alleviation. *Econ. Polit. Wkly*. **2003,** *38* (5), 432–433. http://www.jstor.org/stable/4413155

Karim, L. Analyzing Women's Empowerment: Microfinance and Garment Labor in Bangladesh. *Fletcher Forum World Aff.* **2014,** *38* (2), 153–166. http://www.jstor.org/stable/45289749

Khan, S. I.; Hussain, M. I.; Parveen, S.; Bhuiyan, M. I.; Gourab, G.; Sarker, G. F.; Arafat, S. M.; Sikder, J. Living on the Extreme Margin: Social Exclusion of the Transgender Population (Hijra) in Bangladesh. *J. Health Population Nutr.* **2009,** *27* (4), 441–451. http://www.jstor.org/stable/23499634

Khandakar Qudrat-I Elahi. Microfinance, Empowerment, and Sudra Women in India. *Dev. Practice* **2003,** *13* (5), 570–572. http://www.jstor.org/stable/4029946

Khandelwal, A. K. Microfinance Development Strategy for India. *Econ. Polit. Wkly.* **2007,** *42* (13), 1127–1135. http://www.jstor.org/stable/4419411

Khandker, S. R.; Samad, H. A. Microfinance Growth and Poverty Reduction in Bangladesh: What Does the Longitudinal Data Say? *Bangladesh Dev. Stud.* **2014,** *37* (1 & 2), 127–157. https://www.jstor.org/stable/26538550

Khavul, S. Microfinance: Creating Opportunities for the Poor? *Acad. Manage. Perspect.* **2010,** *24* (3), 58–72. http://www.jstor.org/stable/29764974

Kleynjans, L.; Hudon, M. A Study of Codes of Ethics for Mexican Microfinance Institutions. *J. Busi. Ethics* **2016,** *134* (3), 397–412. http://www.jstor.org/stable/24703779

Leach, F.; Shashikala S. Microfinance and Women's Empowerment: A Lesson from India. *Dev. Practice* **2002,** *12* (5), 575–588. http://www.jstor.org/stable/4029403

M. P. Vasimalai,; Narender. K. Microfinance for Poverty Reduction: The Kalanjiam Way. *Econ. Polit. Wkly*. **2007,** *42* (13), 1190–1195. http://www.jstor.org/stable/4419419

Menon, N. Sexuality, Caste, Governmentality: Contests Over "Gender" in India. *Feminist Rev.* **2009,** *91*, 94–112. http://www.jstor.org/stable/40663982

Mensah, E. A.; Boakye, K. A. Conceptualizing Post-COVID 19 Tourism Recovery: A Three-Step Framework. *Tour. Plan. Dev.* **2021,** 1–25.

Milgram, B. L. Operationalizing Microfinance: Women and Craftwork in Ifugao, Upland Philippines. *Human Organiz.* **2001,** *60* (3), 212–224. http://www.jstor.org/stable/44127259

Mohamed, K. S.; Temu, A. E. Access to Credit and Its Effect on the Adoption of Agricultural Technologies: The Case of Zanzibar. *Afr. Rev. Money Finance Bank.* **2008,** 45–89. http://www.jstor.org/stable/41410533

NABARD. Stastistics, 2021. www.nabard.org/contentsearch.aspx?AID=292&Key=https%2525253a%2525252f%2525252fwww.nabard.org%2525252fauth%2525252fwritereaddata%2525252ftender%2525252f1207215910SoMFI%25252b2020–21.pdf

Pantelić, A. A Comparative Analysis of Microfinance and Conditional Cash Transfers in Latin America. *Dev. Practice* **2011,** *21* (6), 790–805. http://www.jstor.org/stable/23048402

Pitt, M. M.; Khandker, S. R.; Cartwright, J. Empowering Women with Micro Finance: Evidence from Bangladesh. *Econ. Dev. Cult. Change* **2006,** *54* (4), 791–831. https://doi.org/10.1086/503580

Power, A. Social Exclusion. *RSA J.* **2000,** *148* (5493), 46–51. http://www.jstor.org/stable/41378926

Razzaque, M. A. Microfinance and Poverty Reduction: Evidence from a Longitudinal Household Panel Database. *Bangladesh Dev. Stud.* **2010,** *33* (3), 47–68. http://www.jstor.org/stable/23339852

Rebecca M. V., Schreiner, M. Women, Microfinance, and Savings: Lessons and Proposals. *Dev. Practice* **2002,** *12* (5), 602–612. http://www.jstor.org/stable/4029405

Satish, P. Mainstreaming of Indian Microfinance. *Econ. Polit. Wkly.* **2005,** *40* (17), 1731–1739. http://www.jstor.org/stable/4416532

Scheyvens, H. Does Microfinance in Its Current Forms Contribute to Adaptation? In *The Role of Microfinance and Microfinance Institutions in Climate Change Adaptation: Learning from Experiences in Bangladesh*; Institute for Global Environmental Strategies, 2015; pp 29–40. http://www.jstor.org/stable/resrep00725.11

Schrieder, G.; Sharma, M. Impact of Finance on Poverty Reduction and Social Capital Formation: A Review and Synthesis of Empirical Evidence/Impacts Positifs De La Microfinance Sur La Réduction De La Pauvreté Et Sur Le Capital Social—Une Revue Et Synthèse De L' évidence Empirique. *Savings Dev.* **1999,** *23* (1), 67–93. http://www.jstor.org/stable/25830680

Sharma, K. Small Loans, Big Dreams: Women and Microcredit in a Globalising Economy. *Econ. Polit. Wkly* **2011,** *46* (43), 58–63. http://www.jstor.org/stable/23047759

Singh, P. Persons with Disabilities and Economic Inequalities in India. *Indian Anthropol.* **2014,** *44* (2), 65–80. http://www.jstor.org/stable/43899390

Sooryamoorthy, R. Microfinance and Women in Kerala: Is Marital Status a Determinant in Savings and Credit-Use? *Sociol. Bull.* **2005,** *54* (1), 59–76. http://www.jstor.org/stable/23620585

Srivastava, N.; Srivastava, R. Women, Work, and Employment Outcomes in Rural India. *Econ. Polit. Wkly* **2010,** *45* (28), 49–63. http://www.jstor.org/stable/40736730

Tavanti, M. Before Microfinance: The Social Value of Microsavings in Vincentian Poverty Reduction. *J. Busi. Ethics* **2013,** *112* (4), 697–706. http://www.jstor.org/stable/23433677

Thorat, S.; Attewell, P. The Legacy of Social Exclusion: A Correspondence Study of Job Discrimination in India. *Econ. Polit. Wkly* **2007,** *42* (41), 4141–4145. http://www.jstor.org/stable/40276548

Zhao, W. The Nature and Roles of Small Tourism Businesses in Poverty Alleviation: Evidence from Guangxi, China. *Asia Pac. J. Tour. Res.* **2009,** *14* (2), 169–182.

Zhao, W.; Ritchie, J. B. Tourism and Poverty Alleviation: An Integrative Research Framework. *Curr. Iss. Tour.* **2007,** *10* (2–3), 119–143.

CHAPTER 10

Strategic Transformation and the New Normal in the Hospitality Industry

PRIYA SINGH[1], SUMEDHA AGARWAL[2], AND ROHAN BHALLA[3]

[1]*Department of Tourism and Hospitality Management, Jamia Millia Islamia, New Delhi, India*

[2]*School of Business Studies, Sharda University, Noida, India*

[3]*Faculty of Management Studies, Jamia Millia Islamia, New Delhi, India*

ABSTRACT

The outbreak of COVID-19 has put human lives at stake in the global economy. The National Restaurant Association of India (NRAI) reports that the country's food services industry, which is valued at 423,865 crores, is responsible for the employment of more than 70,000 individuals. It is concerning that the widespread occurrence of this pandemic has resulted in substantial upheavals in the thriving industry that this sector represents. Because of the epidemic caused by the COVID-19 virus, the travel and tourism industry is on the edge of an existential crisis, which has resulted in an abundance of unanticipated difficulties and roadblocks. The hotel and tourism industries, of which the food business is a crucial component, are doing everything they can to ensure their continued existence. In the backdrop of the threat of the COVID pandemic, emerging trends indicate the strategic changes in the industry's service patterns. The coming years will redefine eating out for the customers as per the new dining normal. The strategic shifts occurring within the hospitality business are the focus of this research study. The intention is to know how the different domains of

Dynamics of the Tourism Industry: Post-Pandemic and Post-Disaster Perspectives and Strategies.
Gül Erkol Bayram and Anukrati Sharma (Eds.)

hospitality will adapt to the changing paradigm. This chapter has discussed the overwhelming reservations of the customers toward hospitality services. This study contributes to a deeper comprehension of the new management techniques that have taken place within the tourist and hospitality business as a whole and within the food and beverage sector in particular. The transformed management practices will help capture the altered strategies of the industry, thus addressing the customers' expectations of the new normal. This chapter offers a post-pandemic playbook regarding the impending shifts in the standards of the hospitality business as well as the new normal. This article has discussed the revival of the hospitality industry post-COVID-19 from the supply side, addressing the apprehensions from the demand side. The study has derived some meaningful strategies and the reformed protocols adopted by the industry to meet the customers' expectations and changing preferences.

10.1 INTRODUCTION

The tourism and hospitality industry is susceptible to various risks, including epidemics, natural disasters, and disasters caused by humans. The occurrence of disasters has significantly impacted the tourism and hotel industry. The experts have taken various measures to overcome the challenges caused by the crisis. For instance, in the wake of the terrorist events of September 11, 2001, hotels in Hong Kong improved their security systems by installing the most recent generation of closed-circuit television systems and provided their staff with intensive safety education (Chan and Lam, 2013). Following the emergence of severe acute respiratory syndrome (SARS), hotels in Korea made substantial investments in cutting-edge hygienic technology and educated their employees on the need to maintain good health (Kim et al., 2005). During the devastation caused by the earthquake and tsunami in 2011, the hotels located in the coastal regions of Japan provided shelter and protection to evacuees (Nguyen et al., 2017). This was done to accommodate the large number of people who were forced to flee their homes. It is evident from these situations that every situation is unique and demands actions as per the situation. Thus, the hospitality industry needs to address the changes caused due to the unprecedented effects of COVID-19 and try to regain the customers' trust and deliver credible services. This chapter looks at how the COVID-19 pandemic has impacted the hotel sector and discusses some possible responses to the issues that have arisen as a result of the pandemic.

10.2 COVID-19 AND HOSPITALITY

Similar outbreaks like COVID-19 happened earlier, like SARS coronavirus in the early 2000s. The SARS outbreak caused the first known instance of a global health crisis. The World Health Organization (WHO) listed Guangzhou and Hong Kong as high-risk tourist destinations in November 2002, after the first severe acute respiratory syndrome (SARS) cases were discovered in Guangdong Province in southern China. China has multiple regions that have been designated as epidemic zones. In the case of the new coronavirus, a situation very similar to this one emerged. The outbreak of the COVID-19 virus was recognized as a worldwide pandemic and a public health emergency of international importance by the World Health Organization (WHO) on January 30, 2020 (Statement, 2020). There is a link between the viruses since it is highly likely that they both followed a zoonotic (transmission from animals to humans) incident. A global crisis like disease outbreaks and pandemics calls for readiness and crisis management. The problem brought about a decline in the economy and brought up concerns of unfairness since the burdens of such pandemic outbreaks are frequently shouldered by the less fortunate members of society and the more vulnerable locations. At the crucial time, countries across the globe assessed the impacts of COVID-19 on both monetary and nonmonetary scales. After spreading diseases such as Ebola, the Zika virus, and HIV/AIDS, as well as the social backlash that accompanied these events, some regions have had a complex time building resilience (Novelli et al., 2018).

The hotel business has sustained substantial damage due to the epidemic, and it will be quite some time before it can fully recover. Some techniques that have been implemented to reduce the size of the COVID-19 curve include lockdowns, increasing the physical distance between buildings, instituting Janta curfews, and placing limits on movement and mobility. Because of these measures, several hotel industry businesses have had to close their doors temporarily. There was a substantial decline in total demand for the services granted permission to operate (Bartik et al., 2020). Take-out orders and delivery services were the only services that the restaurant provided. The relevant authorities imposed travel restrictions and stay-at-home orders, which led to a fall in the number of hotel bookings, resulting in low or no earnings for the hotels. During the epidemic, the desire for social separation brought on by the dread of the COVID-19 virus had a devastating impact on the restaurant business (Gössling et al., 2020).

The tourism and hospitality sector was brought to its knees by the international health crisis that COVID-19 created, and ever since then, the industry has been trying to stay in business (Chan and Lam, 2013). The decline in tourism and travel has had a negative impact on hotel occupancy rates, which has contributed to a downturn in overall economic activity (Hoisington, 2020). With the global cancellations and postponement of the reservations, the revenue scale of hotels witnessed a colossal fall. The COVID-19 pandemic severely affected hotel operations around the world. The reopening process eventually began at the beginning of 2021, and the restrictions were relaxed, like allowing the dine-in restaurants to operate with minimum seating to ensure physical distancing. But the second wave of the pandemic broke the economy in the worst manner. The industry had been dealing with the COVID-19 pandemic for some time, and the second wave proved disastrous. The ongoing pandemic from 2019 to 2021 had devastating effects on the operations, employees, and customers. Despite all the negatives, there were some silver linings too. The COVID-19 situation resulted in a phase in which novel ideas were explored. The pandemic provided some great opportunities for the hospitality industry. The hospitality industry enhanced its preexisting strategies and underwent a strategic transformation of its operations to accommodate its clients' developing requirements and exhibit the industry's new standard.

10.3 STRATEGIC TRANSFORMATIONS POST-PANDEMIC

The hospitality industry is known for creating a memorable stay for the guests. Exceeding the expectation and ensuring an overwhelming experience for the guest has been a paramount industry goal so far. There have been enough discussions on physical distancing, mask etiquettes, and hygiene practices adopted by the industry post-pandemic. Still, the most critical question was to regain the guests' trust and offer them a credible hospitality experience. The hospitality experts suggested the following strategies to ensure satisfactory experiences for the guests.

10.3.1 ANTICIPATING THE EXPECTATIONS OF THE STRESSFUL GUEST

The pandemic has caused extreme panic among people, and they have become very apprehensive about the quality of services they would receive while eating out. Every guest visiting the hotel is anxious and has a dilemma of being safe on the property. From when they leave home till the time they reach back, their mind is stressed about frequent sanitising of hands, maintaining physical distancing, and wearing masks on their face all the time. The pandemic has been the most crucial time for the industry to regain the trust of the consumers and not to violate the trust with faulty processes and services. The industry should prepare itself well to cope with anxious guests and calm them by winning their trust of ensuring them a perfect and hassle-free stay in the hotel. This unfortunate time of COVID-19 has allowed the industry to rebrand itself and create a positive image in the market. The industry can only survive if it overcomes the upcoming challenges of the sector and embraces the new normal.

10.3.2 IMPLEMENTATION OF CREATIVE MARKETING CAMPAIGNS

Consumers have always relied on the reviews posted by the other consumers as it helps them know about the good and the inadequate services. During the crisis brought on by COVID-19, the hospitality industry must reach out to its clients and focus on their individual experiences. It is of utmost importance for hotels to follow up with guests during and after their stay to obtain their feedback. The hotel personnel should try to extend exceptional services and amenities to guests during their stay. Extraordinary services will open the avenues for the hotels to attract more guests in the future as one satisfied guest shall share their experiences with a few more prospective guests. Hotels should communicate more often with the guests as compared to the pre-COVID-19 times. An example of a creative marketing plan used to understand the perception of people and explain to them about the services of the hotel is illustrated through the case of OYO Hotels and Homes in Box 10.1.

10.3.3 INTRODUCING THE NEW WAYS OF GUEST COMMUNICATION

The way the industry needs to connect with guests has changed dramatically. Traditionally, the emphasis was on direct communication with the guests. The commencement of the COVID-19 pandemic necessitated the maintenance of physical distance, which resulted in the creation of new communication routes. Mobile check-in has been encouraged for a decade now. New touchpoints have been created to facilitate communication

Box 10.1—Creative Marketing Campaign—OYO

The COVID-19 crisis was disastrous for the whole hotel industry. The renowned Indian multinational brand of leased and franchised hotels, houses, and living spaces, OYO Hotels & Houses, was no exception. Mayur Hola, Head of Global Brand for OYO Hotels & Homes, conceived the video campaign Fir Badhega India to demonstrate their uncompromising entrepreneurial spirit and express their preparation for the new standard of living. The month-long campaign highlighted OYO's Sanitised Stays programme and honored the optimism and fortitude of India's 1.3 billion residents as they prepare to move forward. This programme has primarily emphasized the preventative measures taken by the hotel chain to preserve cleanliness, hygiene, and minimal-touch standard operating procedures so as to safely welcome guests. The commercial highlights the new standard of living. The video features an elderly merchant who attempts to reopen his store after it has been shuttered for months. Soon, a young bystander interrupts his hectic schedule to assist the elderly shopkeeper in re-raising the shutter. They exchange pleasantries, and then the good Samaritan immediately hurries off to work. Both protagonists adhere to health and safety measures for the course of the film. The young staff executive also observes OYO's minimal touch experiences, such as temperature checks upon entering the hotel, opening the door with his shoulder, the use of protective equipment such as masks and gloves, the placement of hand sanitisers at high touchpoints, and the mention of OYO's 10-step room sanitization process.

with the guest and resolve their query at various stages during their stay. Text-based messaging over WhatsApp has been the latest trend these days. Automated voice assistants and artificial intelligence have taken over the traditional interactions between guests and operators to encourage socially distant communication. The ongoing pandemic has brought a drastic change in the preferences of the guests also. It has urged the industry to adjust to the new normal ways of communicating with the guests and meeting their desired needs and expectations. Digital communication has facilitated the industry with a robust communication strategy to tap into the potential guests and ensure they come back to the hotels. Even the food and beverage sector has digitally transformed its menu and order-taking procedures (Rodríguez-Antón and Alonso-Almeida, 2020). The industry is trying hard to cope with the challenges and set up a new normal communication technique.

10.3.4 A SURGE IN FOOD DELIVERY SERVICES

Ancillary revenues of the hotel industry have faced a significant fall, and for quite some time, the primary source of income will be the room sales only. Consumers are refraining from using the facilities of spas or other luxuries like the souvenir shops, bars etc. They are even restricting themselves from dining in restaurants. To practice physical distancing, the guests are keener in ordering their food to their doorstep. The hoteliers expect that the room service will make a comeback, so the main focus of the hotels and food and beverage outlets has shifted toward facilitating easy and prompt delivery. Even the luxury hotel chains have entered the market to deliver the food. Many hotels have especially designed delivery menus, and they also market themselves as a trustworthy brand in terms of safety and hygiene.

10.3.5 PREPARING FOR THE FUTURE

The pandemic has compelled the hospitality industry to improve its crisis preparedness. The real challenge for the hospitality industry is to be ready for the upcoming changes in consumer demand and cope with the present issues. The hospitality professionals should increase the preparations and analyze what is expected in the future. The hospitality industry is also attempting to engage and retain their regular customers and rebuild their confidence in the brand. Thorough research needs to be done well in advance to discuss the changing requirements in the operation system. The year 2020 brought a significant setback to the industry, but the learnings from this year will prove a learning platform for the revival of the sector in the coming times. Table 10.1 describes some of the examples of the significant initiatives taken by hotels against COVID-19 worldwide:

TABLE 10.1 Initiatives Taken by Major Hotels Globally Against COVID-19.

Hotel	Marketing Campaign	Key features
Six Senses Hotels Resorts Spas, Thailand	Plan Now, Play Later	Proposed Plan Now, Play Later Certificate Provided an additional 20% incentive that could be applied for room purchases
		Upgrades or purchases on the property, such as meals, spa treatments, and activities, were included in the certificates.
		Offered a safe and hygienic environment
The Betsy Hotel, Florida	The Betsy Was Built for This	Using the slogan "The Betsy Was Built for This," a supplementary page was created (separate from its safety and protocol page)
		Offered two four-story buildings with 130 rooms spanning across an entire city block, surpassing much larger properties.
		Boasts an abundance of outdoor places cooled by a constant Atlantic Ocean breeze.
		Multiple key-activated entries to prevent unwelcome contact between guests.
		Launched promotional video showcasing the open and airy rooms of the hotel so that guests can get a sense of exactly what a stay would be like.
The Hoxton, Chicago	The Floor is Yours	With social distancing in mind, target group travelers (whether family reunions or work outings).
		Guests might reserve a full floor of one of their properties at a discounted fee to play (or work) as hard as they want.
		Increased employee training and awareness
Barceló Hotel Group, Spain	We Care About You	Increased sanitation and disinfection
		Ensured health and safety
		Strengthen the digital experience
		Complied with protocols in food and beverages
		Offered innovative experience
		Special arrangements for business travelers

Source: Adapted from A Betsy Update | The Betsy Hotel South Beach, 2020; Six Senses, 2020; The Hoxton: Open House Hotels in London, Europe and USA, 2020; We Care About You, 2020.

10.4 POST-PANDEMIC FUTURE OF HOSPITALITY

The hospitality industry is gradually recovering from the crisis, but the impacts of the pandemic will take a lot of time to subside. The hospitality industry is preparing to implement significant adjustments to its Standard Operating Procedures (SOPs) in response to the pandemic to prioritize the health and safety of its staff and customers. The modifications would also increase consumers' propensity to patronize well-known brands (Gössling et al., 2020). The industry has survived numerous difficult economic downturns, and the current COVID-19 situation is no different. The complete revival of the hospitality industry may be impending, but the formulation of strategies to ensure transformation needs to be done in advance. Hoteliers need to anticipate and reimagine the new expectations of the guests and should strategies their operations accordingly. A few of the proposed changes in the hospitality industry are as under.

10.4.1 REASSURANCE OF HYGIENE AND CLEANLINESS

Cleanliness and hygiene have always been significant factors for any guest deciding upon the booking of the hotel. In the situation of the pandemic, this has become the leading factor influencing the buying decisions of the guests. Therefore, hotels are expected to upgrade their cleaning standards because the guest would demand reassurance of cleaning and sanitation being done in the hotel. According to the WHO, COVID-19 can be transferred by contacting virus-infected surfaces (World Health Organization, 2020). This calls for the frequent surveillance of hygiene of different surfaces, which are a potential source of disease transmission (Memon et al., 2021). The American Hotel & Lodging Association (AHLA) established the Safe Stay campaign in response to the COVID-19 pandemic, which featured better health and safety practices. AHLA implemented the WHO's Operational Considerations for COVID-19 Management in the Accommodation Sector. The Clean and Safe campaign was created by the Malaysian Association of Hotels (MAH) in an effort to improve the existing cleanliness practices. These up-gradations in the cleaning and hygiene patterns affirm the need for a mandatory hygiene protocol in all countries trying to recover from the crisis. The efficient hygiene practices ensure transparency between the hotel and the guests about reformed cleaning services that meet the new emerging expectations of the prospective guests. Box 10.2 explains the hygiene and cleanliness steps taken by the leading fast-food chain Mcdonald's in times of pandemic. Hoteliers have increased the touchpoints at the premises, which include personal helpdesks along with technology-enabled helpdesks. The advancements in the touchpoint phenomenon will help meet the emerging challenges of cleanliness and safety in hotels. In this regard, it is anticipated that the hotels will implement the improvements listed below:

- Appointment of Cleanliness Manager
- Making the guest aware of the cleaning protocols
- Cleaning product upgradation
- Increase the frequency of disinfection and maintenance.
- Improve food and restaurant safety and sanitation procedures.

10.4.2 ARTIFICIAL INTELLIGENCE AND ROBOTICS

The COVID-19 pandemic showed the importance of incorporating technology into the hospitality industry (Huang and Rust, 2020). The physical distancing regulation restricted individuals from being in close physical proximity to one another, making it the ideal moment to implement robotics technology in the hotel industry. Hotels are likely to eliminate staffed services in favor of contactless services via digitalization. The advent of artificial intelligence and robotics made it possible to serve and pleasure guests in novel ways. These new methods were made possible by the revolution in these fields. A new approach to customer service has emerged as a result of hotels' increasing use of robotics and artificial intelligence (Kuo et al., 2017). During the post COVID times, there would be more complex demands from the guests' end, and therefore efficient tools should be used to deliver the same (Huang and Rust, 2021). It is anticipated that in the future, more and more hotels will start availing robot services to serve their guests. The guests would be able to interact with the robots, and contactless services would be encouraged. As of now, the assistance of robots is limited to a few developed countries only. Artificial intelligence and robotics may take time to evolve in developing countries. Meanwhile, contact-free services should be adopted through the digitalization of menus, brochures, and order taking. An example of the usage of robots in service delivery is presented in Table 10.2.

Box 10.2—Safety First: McDonald's India Goes Contactless

COVID-19 had bought significant disruptions in the QSRs of India, and McDonald's India was no exception. Westlife Development, which owns and runs McDonald's outlets in Western and Southern India, became the first quick-service restaurant (QSR) in the world to provide contactless delivery during the staged recovery from the COVID-19 lockdowns. Smita Jatia, Director, Westlife Development, commented on the statement, "Our clients' safety has been and will continue to be our first priority." We are providing our consumers with a handy option to receive safe and hygienic food when they walk out to get necessities by launching this contactless take-out service. This initiative focused on ensuring safe and hygienic options for the customers. The idea of contactless ensured social distancing at each step of service; the customers could conveniently make orders through McDelivery App, make digital payments and collect the orders from the takeaway counters. This development strengthened the safety and hygiene throughout the operations for both the employees as well as the guests. McDonald's India is well known for its safety and hygiene processes, and the brand has left no stone unturned to ensure more stringent hygiene practices during the crisis. In the wake of the pandemic, the company also created the special designation of Chief Safety Officer', which was offered to different team members on a rotation basis. The main job of this role was to ensure the adequate standards of safety and hygiene being followed by the employees of the store at any given point in time.

Amit Jatiya, Vice Chairman, Westlife Development, said, Crisis is the true test of the leader. I have always believed that all times are good times for business, provided you can quickly realign strategies to manage the situation at hand and emerge stronger from it.

Source: Hospitality News (2020).

TABLE 10.2 Use of Robotics and Technology.

Hotel	Technology Used	Key features
Robot Restaurant, Bangalore	Android Robots	Android robots meet and greet the guest and escort them to their tables. Every table has a digital tablet for the guest to choose the dish from the menu and order. A team of six robots serves the food to the guests. Robots are also programmed to sing songs on special occasions.
Robo Chef, Orissa	Humanoid Robots SLAM (Simultaneous Localisation And Mapping)	The robots have 17 kinds of sensors that can sense the environment, heat, and smoke. Humanoid robots welcome guests and take orders from them. Robots continuously work for 8 h. The engagement of humans is also seen to give a human touch.
FOODOM Tianjiang Food Kingdom, China	Artificial Intelligence and Robots	The restaurant deploys robot chefs and waiters to serve food. Robots ensure complete hygiene and no human contact. The restaurant deploys 40 robots that attend to the guests and serve the food.
Dawn Avatar Robot Café, Japan	Humanoid Robots- The OriHime-D robots	Robots are equipped with a camera, microphone, and speakers to speak and receive orders. Robots are controlled through the eye movements of the guest. Guests feel amazed to have such a technology-driven experience.

Source: Adapted from TOI (2020).

10.4.3 ENSURING HEALTH AND SAFETY

Every industry should place a high priority on ensuring the health and safety of their workforce as well as the people they serve in the marketplace. The hospitality industry should give undivided attention to the safety of its stake-holders. Regular and compulsory screening for COVID-19 symptoms should be a routine practice for hotels. There should be strict norms for maintaining physical distance between guests and employees. Protective clothing like PPE kits and masks should be the new normal uniform for the employees that are taking care of the needs of the guests. Hotels should not hesitate to inquire about their guests' recent travel history and prohibit those guests who have traveled to high-risk zones.

Amid the COVID-19 pandemic, lifestyle management has captured a lot of attention (Wang et al., 2020) (Wang et al., 2020). People have started focusing on their physical and mental well-being. Facilitating the wellness needs of the guests should become a new trend for the hotels post-pandemic. The most up-to-date trend in hospitality is providing guests with opportunities to participate in health and wellness activities such as meditations, diet programmes, detox programmes, and sleep hygiene programmes. Although the hospitality industry is frequently witnessing new directions, a lot needs to be done to enhance the facilities required for catering for the guests. Guests' behavioral data and their previous consumption patterns will enable the hotels to develop customised health-centric programs for the guests.

Because of the global issue of COVID-19, travelers will be more concerned about safety and health facilities when making trip decisions. Therefore, the marketing content should incorporate and highlight the unique features and preparedness of the hotel to ensure the secure stay of the guest. The assurance from the hotels will guarantee the guest a sense of security during their visit. A very creative approach for explaining to guests the preparedness of the hotel against COVID-19 presented by Hotel San Luis Obispo in Central California is described in Box 10.3.

10.4.4 RIGOROUS TRAINING OF EMPLOYEES

The outbreak of coronavirus serves as a catalyst for innovations and transformation. With the upcoming technological advancement and digitalization in the processes, there should be rigorous training schedules for the employees. The hotels should hold frequent meetings to discuss the action plans and coping strategies to combat the impacts of COVID-19. The coming years present an opportunity for the industry to reinvent and strategically transform its processes to cater to the needs of the post-pandemic world. Both industry practitioners and customers must learn to accept the sector's new normal.

10.5 POST-PANDEMIC PLAYBOOK

The hospitality industry's revival may be delayed, but it is necessary. Although the recovery will take a longer time as compared to other sectors. The new

Box 10.3 "COVID-19 Prevention Plan" by Hotel San Luis Obispo—California

As things started to get back to normal across the world, the crew at Hotel San Luis Obispo, a new boutique hotel in Central California, was busily working to implement new rules for careful sanitation. They developed a variety of indicators by drawing on the direction and information provided by local health authorities as well as organizations within the government. There was a whole page on the website devoted to the hotel's rules and regulations for safety and security. The hotel gave a humorous presentation of the procedures that needed to be followed. They did not present the procedures in the form of bullet points followed by prose as many other hotels do; rather, they drew creativity from the playful brand approach they take. They came up with a graphical technique to demonstrate their precautions and procedures, which they used. The content was engaging, maintained the image of the business throughout, and most importantly, it provided useful information. The following subjects were covered in depth within the material of the website:

- *Cleaning and Sanitation*
- *Temperature and Wellbeing Check*
- *Use of Hand Hygiene, Masks, and Gloves*
- *Social Distancing*
- *Rules of Guest Rooms*
- *Breakfast Options in the Lobby*
- *Dining and Guest Requests*
- *Amenities and Services*

The campaign caught the eye of the guests and was an immediate success because of the graphical representation of content

Source: Hotel San Luis Obispo.

normal will witness a drastic paradigm shift, and a lot of changes are proposed by industry experts in the wake of this pandemic. The post-pandemic playbook acknowledges the ongoing damage to the food and beverage industry and suggests remedial measures. Its goal is to help the food and beverage industry to revive through developing strategies around the ramifications of the COVID-19 pandemic. These inevitable transformations help in designing the post-pandemic playbook, as explained in Figure 10.1:

10.6 UNDERSTANDING THE NEW NORMAL OF THE FOOD AND BEVERAGE INDUSTRY

During the pandemic period, the food and beverage industry underwent fast digitalization, and it is highly unlikely that this progression will slow down in the foreseeable future. Businesses are embracing new technology to improve their processes and level of competitiveness in a world that is increasingly controlled by automated processes, intelligent systems, and Internet of Things (IoT) devices (Telukdarie et al., 2020). Artificial intelligence has been demonstrated to be useful in a variety of ways within the food business. In recent years, the day-to-day operations of the Indian food company have been successfully integrated with cutting-edge technology. Companies that rely on more conventional approaches to customer service have a distinct technological disadvantage when compared to those that employ artificial intelligence since it gives them a clear competitive advantage. Companies that deliver food, such as Zomato, Swiggy, and Foodpanda, have adopted artificial intelligence (AI) to provide their customers with the best possible experience. In addition, a significant number of restaurants are developing their own logistic systems based on artificial intelligence to better serve their clients and provide superior experiences (Kumar, 2020).

In contrast, India's standards on the safety of its food and its level of cleanliness have increased dramatically in recent years. The food and beverage industry has recently adopted a number of new requirements, some of which include the regulation of temperatures, the maintenance of a regular cleaning regimen, the utilization of secure packaging methods, and the wearing of face masks. In addition, health and safety auditing organizations are used on a regular basis to ensure a tight follow-up on the hygiene practices, which further increases the customer's faith in the company. This increases the customer's trust in the company. The Food Safety and Standards Administration International (FSSAI) has issued a

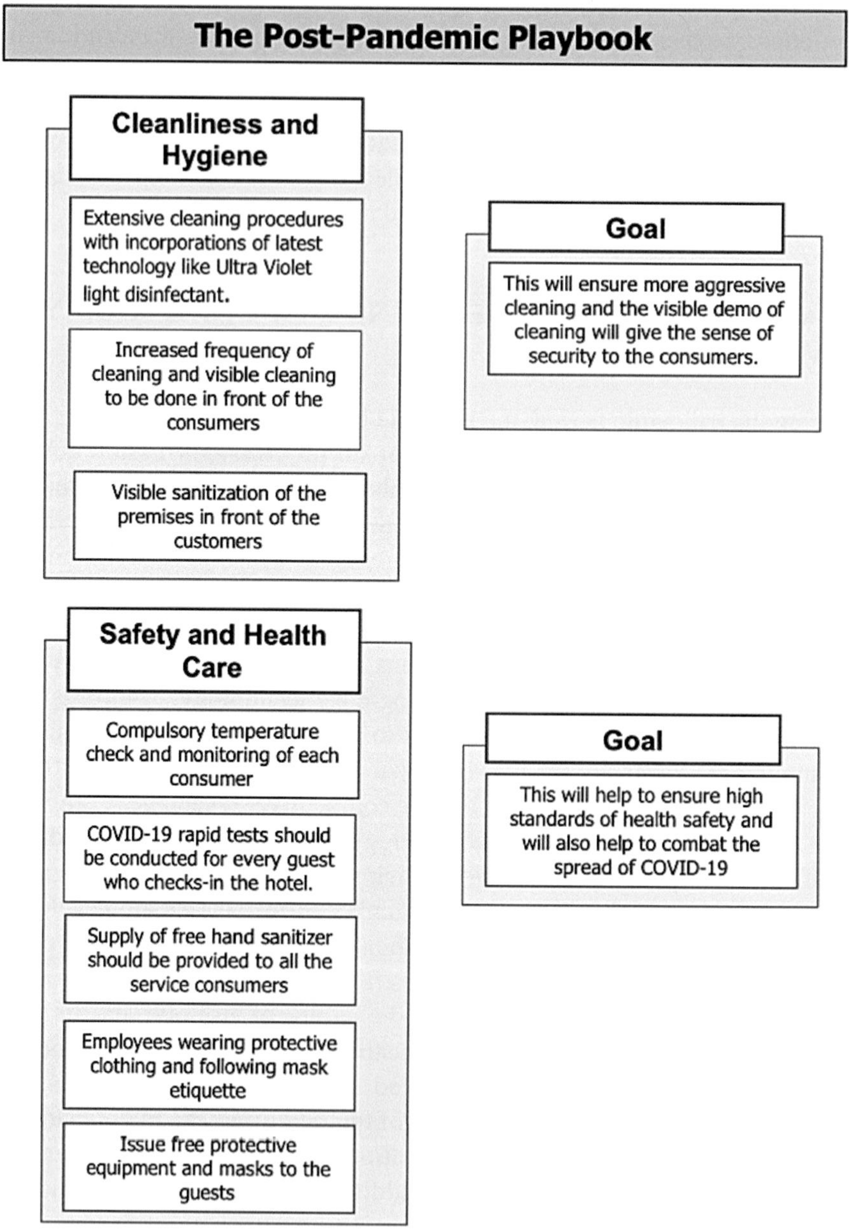

The Post-Pandemic Playbook

Cleanliness and Hygiene

- Extensive cleaning procedures with incorporations of latest technology like Ultra Violet light disinfectant.
- Increased frequency of cleaning and visible cleaning to be done in front of the consumers
- Visible sanitization of the premises in front of the customers

Goal

This will ensure more aggressive cleaning and the visible demo of cleaning will give the sense of security to the consumers.

Safety and Health Care

- Compulsory temperature check and monitoring of each consumer
- COVID-19 rapid tests should be conducted for every guest who checks-in the hotel.
- Supply of free hand sanitizer should be provided to all the service consumers
- Employees wearing protective clothing and following mask etiquette
- Issue free protective equipment and masks to the guests

Goal

This will help to ensure high standards of health safety and will also help to combat the spread of COVID-19

(Continued)

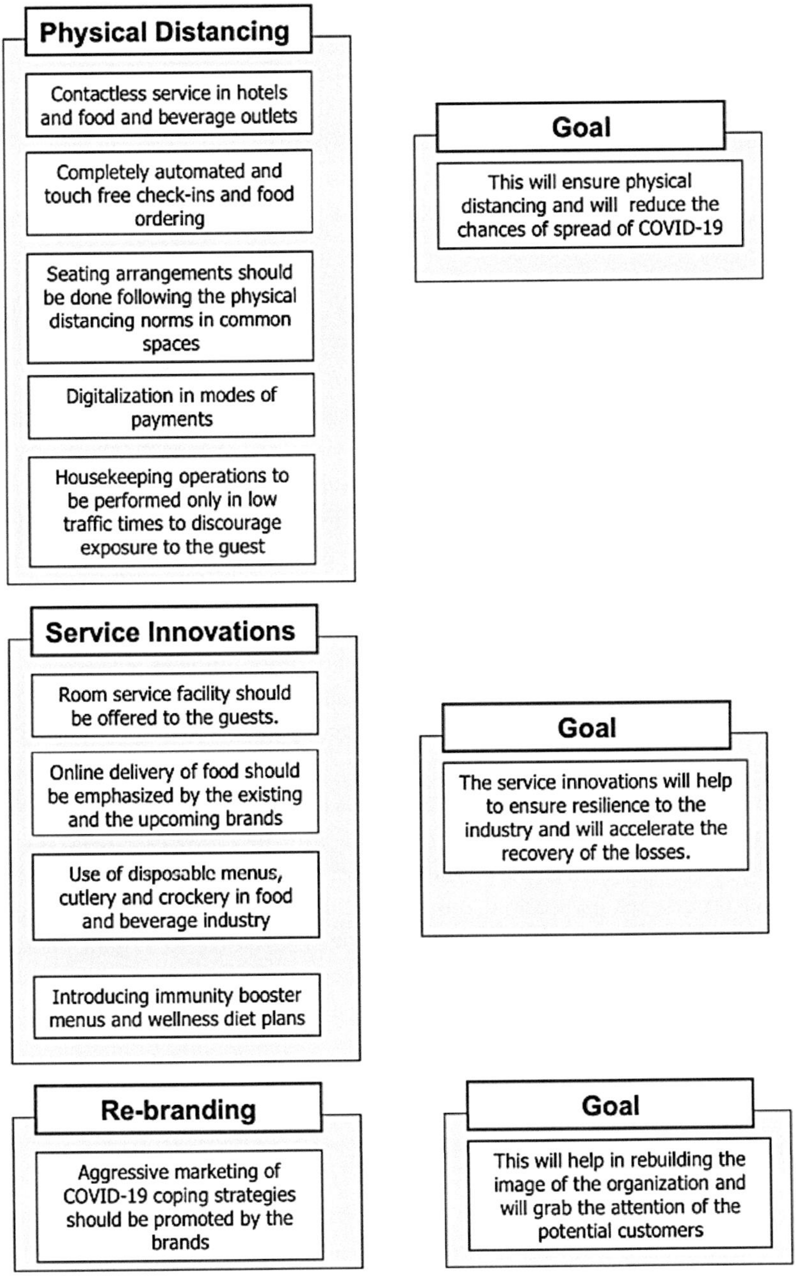

Physical Distancing

- Contactless service in hotels and food and beverage outlets
- Completely automated and touch free check-ins and food ordering
- Seating arrangements should be done following the physical distancing norms in common spaces
- Digitalization in modes of payments
- Housekeeping operations to be performed only in low traffic times to discourage exposure to the guest

Goal

This will ensure physical distancing and will reduce the chances of spread of COVID-19

Service Innovations

- Room service facility should be offered to the guests.
- Online delivery of food should be emphasized by the existing and the upcoming brands
- Use of disposable menus, cutlery and crockery in food and beverage industry
- Introducing immunity booster menus and wellness diet plans

Goal

The service innovations will help to ensure resilience to the industry and will accelerate the recovery of the losses.

Re-branding

- Aggressive marketing of COVID-19 coping strategies should be promoted by the brands

Goal

This will help in rebuilding the image of the organization and will grab the attention of the potential customers

FIGURE 10.1 Post-pandemic playbook.

Source: Authors' interpretation.

48-point checklist for restaurants to implement regarding hygiene ratings. This was done with the intention of eventually generating more strict hygiene standards and safety auditing procedures (Consulting, 2020). The extent to which consumers relied on internet platforms to fulfil their requirements were significant. Because of this, the food and beverage industry underwent a variety of various types of reorganization. The business sector as a whole is undergoing a profound transformation, most noticeably in the following areas:

10.6.1 DINING OUT EXPECTATIONS AND EXPERIENCES

During the early stages of 2020, the industry witnessed a considerable fall of approximately 80% in the business. However, later in 2021, the easing restrictions and increased vaccinations have led to a remarkable recovery in the food and beverage market globally. Furthermore, the role of delivery partners and food aggregators has immensely contributed to the revival of the industry. Across the world, food delivery services have made it possible for consumers to remain connected with the food outlets, which helped in the sustenance and revitalization of the food industry (Reply, 2020). Furthermore, amid the pandemic, consumers have been practising the trend of Eating in by Ordering Out, which certainly added to the income of the food and beverage industry.

Additionally, the ongoing drop in COVID-19 incidence and break-throughs in vaccination has led to a global shift in consumer behavior. The diners have started revisiting the cafes, restaurants, and bars. Nevertheless, the consumers are aware and considerate of the hygiene protocols being followed by the food and beverage sector. Therefore, to regain customer trust and offer a credible dining experience to guests, the industry is striving hard to incorporate new operational strategies and ensure COVID hygiene protocols.

10.6.2 TECH-LED RECOVERY FOR FOOD AND BEVERAGE SECTOR

Before COVID-19, the services offered by the food and beverage sector were based on the traditional system of guest interaction. There was face-to-face contact with customers, and their need was closely analyzed. The pandemic resulted in strict social distancing, and the entire process went online. The pandemic crisis compelled the industry to think back for some innovative ideas to regain market momentum. The fear of the spread of the disease called for contactless guest engagement. During the COVID-19 era, touchless solutions and digital payments became the new norm in the hospitality industry (IBEF, 2020). The implementation of contactless solutions and self-service stations, as well as an increased emphasis on health, safety, and protection, has become the standard not only in India but all around the world. Robots that deliver food are no longer just a dream for the industry. Some of the most common technological changes that can be seen in the industry today are digital menus and easy-to-use in-app ordering. Moreover, innovations like App-ordering and Digital-Payments' are likely to continue post-crisis as well. The tech-driven solutions of the food and drink industry have been quickly adopted by consumers. From checking in to placing an order to making a payment, everything seems to be done digitally. For example, the hygiene reasons at restaurants and various sales points have caused a noticeable growth in Quick Response (QR) technology. Now, the food and beverage eating experience will include an online pre- and post-experience.

10.7 CHALLENGES ARISING FROM THE NEW NORMAL OF THE HOSPITALITY INDUSTRY

During the unprecedented pandemic times, the strategic transformations within the industry have undoubtedly led to its revival, although the challenges arising from the developments are inevitable. The hospitality industry has always been known for its human touch and guest engagements. With the digitalization and the robotic intervention in the food and beverage sector, the charm of the foodservice seems to fade. Before COVID-19, the face-to-face interactions among the guests and the employees accelerated in delighting the customer's experience. Contactless solutions have made it hard for consumers today to express their special requests or customised services. The food and beverage industry has continually appraised its employees through huge tips and incentives. The digital payment mode has left no or minimum choices for the guests to pay tips to the servers. With the exciting coupon codes and heavy discounts offered by the food delivery apps, there has been a tremendous increase in online food orders. Still, it does not replicate the experience of dining out. The means to overcome the shortcomings mentioned above is debatable and yet to be researched and answered.

10.8 CONCLUSION

The hospitality industry has faced a lot of disruptions due to the COVID-19 outbreak. Now is the time for the revival of the sector. Careful examination of the crisis will undoubtedly enhance the industry practices. Alterations to the management techniques are required as a result of the transformations that have taken place in the surrounding environment. The industry has been seen welcoming all the new reforms and changes to combat the COVID-19 pandemic.

To some extent, this global pandemic has revived the hospitality industry globally. Amid the pandemic, there have been a few remarkable changes in the industry like digitalization of processes, contactless and prompt service, adoption of new marketing strategies, readiness for crisis management etc. Soon, the industry will make strides in improving the mental and physical health of consumers. The pandemic has also affected the sentiments of the consumers. Consumers have new preferences and expectations from the industry now. The need for safety and security now replaces luxury. Customers would now be very particular about their selection of services and look for their safety as a prime concern. The industry and consumers both are looking forward to welcoming the new normal of hospitality.

KEYWORDS

- **COVID-19**
- **new normal**
- **strategic transformation**
- **food and beverage industry**

REFERENCES

A Betsy Update | The Betsy Hotel South Beach, 2020. https://www.thebetsyhotel.com/a-betsy-update

Bartik, A.; Cullen, Z.; Bertrand, M.; Glaeser, E. L.; Luca, M.; Stanton, C. How Are Small Businesses Adjusting to COVID-19? Early Evidence from a Survey. *SSRN Electron. J.* **2020,** https://doi.org/10.2139/ssrn.3570896

Chan, E. S. W.; Lam, D. Hotel Safety and Security Systems: Bridging the Gap Between Managers and Guests. *Int. J. Hosp. Manage.* **2013,** *32* (1), 202–216. https://doi.org/10.1016/j.ijhm.2012.05.010

Consulting, D. *China's Food Delivery Market Dynamics*; 2020.

Gössling, S.; Scott, D.; Hall, C. M. Pandemics, Tourism and Global Change: A Rapid Assessment of COVID-19. *J. Sustain. Tour.* **2020,** *0* (0), 1–20. https://doi.org/10.1080/09669582.2020.1758708

Hashmi, A. *Fir Badhega India Archives—Official OYO Blog*. https://www.oyorooms.com/officialoyoblog/tag/fir-badhega-india, (Accessed 1 July 2020).

Hoisington, A. *5 Insights About How the COVID-19 Pandemic Will Affect Hotels | Hotel Management*. https://www.hotelmanagement.net/own/roundup-5-insights-about-how-COVID-19-pandemic-will-affect-hotels.

Hospitality News. *McDonald's Goes Contactless for the Sake of Safety and Hygiene*. ET HospitalityWorld, 2020. https://hospitality.economictimes.indiatimes.com/news/restaurants/mcdonalds-goes-contactless-for-the-sake-of-safety-and-hygiene/76026846, (Accessed 1 July 2020).

Hotel San Luis Obispo. *Health + Safety*, 2020, https://hotel-slo.com/hotel/safe-travels/health-and-safety/, (Accessed 24 Feb 2022).

Huang, M. H.; Rust, R. T. Engaged to a Robot? The Role of AI in Service. *J. Serv. Res.* **2021,** *24* (1), 30–41. https://doi.org/10.1177/1094670520902266

IBEF. *Future of Indian Food and Beverage Industry Post-Pandemic*, 2020.

Kim, S. S.; Chun, H.; Lee, H. The Effects of SARS on the Korean Hotel Industry and Measures to Overcome the Crisis: A Case Study of Six Korean Five-Star Hotels. *Asia Pac. J. Tour. Res.* **2005,** *10* (4), 369–377. https://doi.org/10.1080/10941660500363694

Kumar, S. *Amalgamation of AI in Food Tech Industry of India*. Startuptalky, 2020. https://startuptalky.com/amalgamation-of-ai-ml-in-indias-food-tech-domain/

Kuo, C. M.; Chen, L. C.; Tseng, C. Y. Investigating an Innovative Service with Hospitality Robots. *Int. J. Contemp. Hosp. Manag.* **2017,** *29* (5), 1305–1321. https://doi.org/10.1108/IJCHM-08-2015-0414

Memon, S. U. R.; Pawase, V. R.; Pavase, T. R.; Soomro, M. A. Investigation of COVID-19 Impact on the Food and Beverages Industry: China and India Perspective. *Foods* **2021,** *10* (5), 1069. https://doi.org/10.3390/foods10051069

Nguyen, D. N.; Imamura, F.; Iuchi, K. Public-Private Collaboration for Disaster Risk Management: A Case Study of Hotels in Matsushima, Japan. *Tour. Manage.* **2017,** *61*, 129–140. https://doi.org/10.1016/j.tourman.2017.02.003

Novelli, M.; Gussing Burgess, L.; Jones, A.; Ritchie, B. W. 'No Ebola…Still Doomed'—The Ebola-Induced Tourism Crisis. *Ann. Tour. Res.* **2018,** *70*, 76–87. https://doi.org/10.1016/j.annals.2018.03.006

Reply. *A New Push for Digitalization in the Food and Beverage …* , 2000. https://www.reply.com/en/content/a-new-push-for-digitalization-in-the-food-and-beverage-industry

Rodríguez-Antón, J. M.; Alonso-Almeida, M. D. M. COVID-19 Impacts and Recovery Strategies: The Case of the Hospitality Industry in Spain. *Sustainability (Switzerland)* **2020,** *12* (20), 1–17. https://doi.org/10.3390/su12208599

Six Senses. 2020. https://www.sixsenses.com/en/offers/plan-now-play-later

Statement, W. *International Health Regulations (2005) Emergency Committee Regarding the Outbreak of Novel Coronavirus (2019-nCoV)*, 2020 www.who.int/news-room/

detail/30-01-2020- statement-on-the-second-meeting-of-the-international-health-regulations-(2005)-emergency-committeeregarding-the-outbreak-of-novel-coronavirus- (2019-ncov).

Telukdarie, A.; Munsamy, M.; Mohlala, P. Analysis of the Impact of COVID-19 on the Food and Beverages Manufacturing Sector. *Sustainability (Switzerland)* **2020,** *12* (22), 1–22. https://doi.org/10.3390/su12229331

The Hoxton. Open House Hotels in London, Europe and USA, 2020. https://thehoxton.com/

TOI. *Bengaluru Just Got Its First Robot Restaurant, Where Robots Will Be at Your Service*. Times of India Travel, 2020. https://timesofindia.indiatimes.com/travel/eating-out/bengaluru-just-got-its-first-robot-restaurant-where-robots-will-be-at-your-service/as70771831.cms

Wang, G.; Zhang, Y.; Zhao, J.; Zhang, J.; Jiang, F. Mitigate the Effects of Home Confinement on Children During the COVID-19 Outbreak. *Lancet* **2020,** *395* (10228), 945–947. https://doi.org/10.1016/S0140-6736 (20)30547-X

We Care About You, 2020. https://www.barcelo.com/en-ww/general-information/notices/we-care-about-you/

World Health Organization. The COVID-19 Risk Communication Package for Healthcare Facilities. In *WHO*; 2020.

CHAPTER 11

Marketing and Competition Strategies of Hotels in the Aftermath of the COVID-19 Pandemic

AHU YAZICI AYYILDIZ[1] and YAVUZ CETIN[2]

[1]*Faculty of Tourism, Kusadasi, Aydin Adnan Menderes University, Aydin, Turkey*

[2]*Institution of Social Sciences, Tourism Management, Aydin Adnan Menderes University, Aydin, Turkey*

ABSTRACT

This chapter explains and discusses the marketing strategies of hotels in the top hotel chains list of Horwath HTL's European Chains & Hotels Report 2019. Based on semistructured interviews with the managers of these hotels, data were collected regarding the main changes that occurred in these hotels as a result of the COVID-19 pandemic regarding the ways of doing business and the target markets and how they responded to these changes by changing their operational, marketing, and competition strategies. The study shows, as a consequence of the COVID-19 pandemic, though the hotels faced major difficulties, by adapting to the changes in terms of changing operation strategies, markets, marketing strategies, and competition strategies, hotel chains have been able to cope with the changes taking place in the environment. It may be stated that the COVID-19 pandemic has made the hotel chains more cost, competition, and marketing-oriented, hence more resilient.

Dynamics of the Tourism Industry: Post-Pandemic and Post-Disaster Perspectives and Strategies.
Gül Erkol Bayram and Anukrati Sharma (Eds.)

11.1 INTRODUCTION

The COVID-19 epidemic which started in Wuhan, China at the end of 2019 and the beginning of 2020 not only resulted in the death of millions of people but also influenced all sectors of the economy. Tourism is one of the most significantly influenced industries by the COVID-19 Pandemic (Lee and Chen, 2021; Kenny and Dutt, 2021; Yiwei et al., 2021). In addition to complete closures in almost all parts of the world, during which almost all business activities, including in tourism and hospitality, came to a halt, yet after the peak of the COVID-19, things have not gone back to normal yet (Han et al., 2021). For instance, in Japan, about 1770 hotels went bankrupt in 2021 alone (Teikoku Databank, 2022).

Tourism is an important industry for Turkey. Tourism constitutes 3.8% of Turkey's GDP and 7.7% of total employment in the country (OECD, 2020). Tourism also accounted for 52% of all service exports in the country (OECD, 2020). As income generated in tourism may influence growth in 30 or so other industries ranging from transportation to food, beverages, furniture, durables, and construction (Koc and Ayyildiz, 2021a), and the fact that tourism multiplier in Turkey is one of the highest in the world (Fletcher, 1995), it is worth investigating the difficulties experienced by the hotel chains in Turkey, and how they changed their marketing strategies in the aftermath of the COVID-19 Pandemic. The competitiveness abilities of hotels, at the micro level, may influence the competitiveness of the tourism and hospitality sector in general, and even hence may have significant implications for the economy at the macro level (Koc, 2009).

11.2 THE INFLUENCE OF COVID-19 ON TOURISM AND HOSPITALITY, AND THE HOTELS

A plethora of studies has been carried out to investigate the influence of COVID-19 on the tourism and hospitality industry. For instance, Sigala's (2021) bibliometric study identified four main themes investigating the influence of COVID-19; i) on the tourism and hospitality stakeholders, ii) managing change and resilience, and transformation, iii) crisis, recovery, and future, iv) on tourist decision-making, destination marketing, and technology adoption. Kim and Han (2022) who compared the importance of hotel selection attributes pre-COVID-19 and post-COVID-19 found major differences between the two. Based on their findings Kim and Han (2022) proposed marketing and competition strategies for hotel managers.

The COVID-19 has had several influences on the hotels. First, with the decline in the occupancy rates, the hotels found it difficult to survive without the revenues coming in (Spanaki et al., 2021). Without the revenues the staff had to be made redundant and/or had to be asked to do overtime, rendering the implementation of human resource management activities extremely difficult (Spanaki et al., 2021). While some hotels and hotel chains went bankrupt, the others took cost reduction and revenue-increasing measures to survive. Yacoub and El Hajjar (2021) showed that in the aftermath of the COVID-19 Pandemic, Lebanese hotel managers switched their attention toward more safety-conscious operations within all departments, placed greater importance on domestic tourists, and increased flexibility in bookings and cancellations to encourage customers. Hidalgo et al.'s (2022) study indicated that hotels in Spain have changed their target market and concentrated more on the MICE (Meeting, Incentive, Conference, and Exhibition) market. In a similar vein, Ayyildiz's (2020a) study in Turkey showed that hotels made adopted turnaround strategies following the COVID-19 pandemic responded to the changes in the environment. Ayyildiz (2020a) argued that hotels became more customer-oriented and marketing-oriented following the pandemic, as the increased levels of competition involved the survival of the fittest.

Smart et al.'s (2021) study in the US showed that hotels had to have new health and safety protocols at the operational level, but they also had to make an important strategic decision. The market segments the hotels relied on for a long period of time appear to have not existed any longer, and hotel managers had to turn to new market segments both geographically and also in terms of the hospitality product they offered.

Hao et. al. (2020) demonstrated that COVID-19 Pandemic has led hotel managers in China to make changes in their product offerings, increase the use of digital media and look for ways to increase membership and loyalty. Herédia-Colaço and Rodrigues' (2021) research in Portugal found that the most strategic measures the hotel managers took were based on special health and safety protocols, and taking marketing initiatives such as long-term vouchers to increase sales. Bonfanti et al.'s (2021) study with international hotel managers showed that managers had to redesign the consumer experience in terms of safety measures to achieve the expected benefits of reassurance, quickness, intimacy, and proximity. For instance, having to wear masks by staff and social distancing reduced the perceptions of intimacy by the customers. Managers with the measures they adopted have been able to create a new type of proximity, emotional and social closeness rather than by physical closeness, through the communications and behaviors of staff. Gursoy et al. (2020) found that more than 70% of customers valued the use of various technologies in service delivery in the COVID-19 environment to reduce personal contact through the use of technologies such as service robots, contactless payment systems, digital menus with QR codes, contactless digital payments, keyless entry, contactless elevators, etc.

Although there are a few examples of studies on how hotel managers have responded to the COVID-19 for recovery, this study by investigating the operational, marketing, and competition strategies of the top ten hotel chains in Turkey in the aftermath of COVID-19 aims to provide a framework for the strategies developed by the hotels in Turkey.

11.3 THE METHOD

As stated above, this study aims to identify the competitive strategies of hotels following the COVID-19 pandemic to understand how they responded to the changes taking place in the market. The top ten hotels in terms of their size in Turkey in the Horwath HTL's European Chains & Hotels Report 2019 were included in the study. The interviews were held with the managers of these hotels, namely Wyndham, Hilton, Accor, Marriott, IHG (InterContinental Hotels Group), Rixos, Anemon, Divan, Dedeman, and Kaya Hotels (Table 11.1).

TABLE 11.1 The Largest Hotel Chains in Turkey.

Rank	Hotel chains	Number of hotels	Number of rooms
1	Wyndham	75	10,953
2	Hilton	63	12,674
3	Accor	42	7453
4	Marriott	30	5686
5	IHG (InterContinental Hotels Group)	27	5026
6	Rixos	26	8721
7	Anemon	18	2210
8	Divan	17	2080
9	Dedeman	17	2784
10	Kaya Hotels	14	5630

Data for the study were collected through semistructured interviews that took place between the 1st and 30th of June 2021, following the prior formal appointments that were set with the managers. The interviews with the managers were held online due to the COVID-19 pandemic precautions. Before the interviews, the literature was reviewed for similar studies and the opinions of one expert academic in the field and one experienced manager in the sector were sought after. For the design of the interviews, the studies of Ayyildiz (2020a), Karadeniz et al. (2020), Herédia-Colaço and Rodrigues (2021), and Bonfanti et al. (2021) were mainly resorted to.

For the analysis of data, the content analysis technique was used. The content analysis can be defined as a scientific method where and which the text materials are analyzed systematically, to be followed by a grouping and categorizing process based on specific criteria to be able to provide a basis for future research (Bowen and Bowen, 2002; Yildirim and Simsek, 2013). Then, the similar responses were grouped under specific themes and tabulated. In doing so the study attempted to ascertain the marketing and competition strategies of the hotel managers in the aftermath of the COVID-19 pandemic. As a result of the content analysis of the responses of the managers to the interview questions, all responses were first evaluated individually.

11.4 THE FINDINGS AND THE ANALYSIS OF DATA

As stated above, the interviews with the hotel managers were held between the 1st and 30th of June 2021. Table 11.2 shows the gender, position, and educational backgrounds of the managers. Table 2 shows that only three of the managers from the top ten hotel chains were women, and they did not hold the positions of general manager. Though women appear to provide and oversee a better service in tourism and hospitality (Koc, 2020a), Table 11.2 shows the signs of vertical and horizontal gender segregation in tourism (Remington and Kitterlin-Lynch, 2018). The above may sound like an inappropriate sphere of discussion in a study of marketing and competition strategies of hotels in the aftermath of COVID-19. However, research shows that company boards with high women representation tend to have a 53% higher return on equity, a 66% higher return on the amount of capital invested, and 42% higher returns on sales (Joy et al., 2007). Moreover, the presence of one female director on the company management board reduces the risk of bankruptcy by 20% (Joy et al., 2007). Furthermore, women tend to be better at multi-tasking, a skill that may be more needed in a time of crisis (Barber and Odean, 2001).

TABLE 11.2 The Participants in the Study.

Participant	Age	Gender	Level of education	Position held
1	43	Male	Undergraduate Degree	General Manager
2	40	Male	Postgraduate Degree	General Manager
3	45	Female	Undergraduate Degree	Sales and Marketing Manager
4	48	Male	Undergraduate Degree	Deputy General Manager
5	39	Female	Undergraduate Degree	Sales and Marketing Coordinator
6	43	Male	Postgraduate Degree	General Manager
7	46	Male	Undergraduate Degree	Deputy General Manager
8	51	Female	Undergraduate Degree	General Manager
9	54	Male	Undergraduate Degree	General Manager
10	50	Male	Undergraduate Degree	General Coordinator

It was shown above that there were about 1770 hotels that went bankrupt in Japan (Teikoku Databank, 2022). In masculine cultures such as Japan and Turkey (Koc, 2020b) there are fewer women in high-status managerial jobs. As a masculine society in Japan, only one in ten top managers and politicians is a woman (Lida, 2018).

The interviews with the managers showed that the main changes the managers experienced were the rising costs, the decline in occupancy rates, and the increase in the number of personnel needed to provide the same service (Table 11.3). These changes while increasing the costs decreased the revenues of the hotels. However, the costs incurred relating to sanitation and hygiene, increased customer satisfaction and trust. This finding is in line with the findings of studies in the literature (Herédia-Colaço and Rodrigues, 2021; Spanaki et al., 2021), showing that customers value the hygiene measures. Similarly, Gursoy et al. (2020) found that 40% of the hotel customers were willing to pay more for increased hygiene and safety precautions.

Table 11.3 also shows the changes in the markets hotels serve. As in the case of Yacoub and El Hajjar's (2021) study in Lebanon, some of the hotel chains in Turkey turned to the domestic market. Furthermore, Table 11.3 demonstrates that the costs rose because additional staff had to be put in place, especially in restaurants, for each buffet. However, instead of the customers serving their food themselves at the buffets, when a staff member serves the food, there may be a decline in the consumption and the wastage of food. Koc (2013) found that open buffet food served on all-inclusive holidays caused an illusion of control and tourists consumed more variety and amount of food and beverages on all-inclusive holidays. All-inclusive

TABLE 11.3 The General Changes that Occurred at the Hotels After the COVID-19.

The main theme	Subthemes	Views of the participants (P)
The Rising of Costs	Sanitation and hygiene costs. The cost of various restaurant equipment The costs of personnel uniforms Certification costs Equipment Costs Advertising and Promotion	**P 1:** *After the COVID-19 pandemic the costs have risen in general. Social distancing signs have been placed all over the hotel from the entrance to all the areas open to guests, in the lifts. Additionally, thermometers have been placed at the entrances.* *Free masks have been provided at the entrances for the hotel guests.* **P 2:** *Disposable hygiene sets have been provided in all of the rooms. We made hotel key cards disposable as well. Though these measures have increased costs, they have been instrumental in increasing customer trust and satisfaction.* **P 7:** *We obtained certificates such as the Safe Tourism Certificate and the Hygiene Certificate which were made compulsory to operate by the Ministry of Culture and Tourism. The guests have placed great importance on the certificates, especially at the beginning of the pandemic. The certificates have increased their trust in the hotel.* **P 9:** *We signed a contract with a good social media agency and turned our attention to digital promotion and advertising.*
The Decline in Occupancy Rates	The decline in demand in the domestic market The decline in demand in the international market The decline in demand in group customers	**P 8:** *The COVID-19 pandemic has influenced extremely negatively the hotels open only for the season due to a ban on international flights. We incurred significant losses. The season became shorter. The occupancy rates we normally have until December saw a sharp decline as of mid-October.* **P 3:** *As the domestic tourists refrained from going on holidays and when they did go on holidays they preferred boutique hotels we faced a major decline in the domestic market.* **P 10:** *Before the COVID-19 we mainly catered for the groups from Anatolia. However, after the ban on flights we lost this market. We had to turn our attention to the domestic market.*
Increase in the Number of Personnel Needed	Having to employ more staff in the restaurants	**P 5:** *As we are a hotel working on an all-inclusive basis, a need arose to employ a separate service staff for each buffet, and provide a mask and a face shield. The queues of guests occurred in front of the buffets, as one staff member had to serve the food. Though there were dissatisfactions and complaints about this procedure, it was something that that to be done during the pandemics.*

holidays not only increased the consumption of food and beverages, but also the wastage of food and beverages served at a hotel (Koc, 2013).

TABLE 11.4 Target Markets of Hotels After the COVID-19.

The main theme	Subthemes	Views of the participants (P)
Changes in the Target Market	Turning to the European Market Turning to the domestic market Turning to the closer markets	**P 1:** *Changes occurred. We were catering for the Far Eastern Market. However, after the COVID-19 pandemic, we turned to the European market and signed contracts with the agencies that worked with the European customers.* **P 3:** *We turned to closer markets that have become important such as Bulgaria, Ukraine, and Belarus. We saw a major decline in the European market.* **P 6:** *First, we turned to the domestic market, and then to the Russian and German markets. The intensive demand by the Turkish customers living in Germany was especially noteworthy.* **P 10**: *When the far eastern market and the group customers market were lost, individual domestic tourists became saviours for us.*
No Change Occurred in the Target	Serving Business Travelers	**P 4:** *As we operate in the business segment we continued to serve in this market. Especially, the UK and Germany form the basis of our customer profile.*

Table 11.4 shows that there have been many changes in the markets the hotel chains served.

As experienced in almost all of the countries studied in the literature (Hao et. al., 2020; Spanaki et al., 2021; Yacoub and El Hajjar, 2021; Smart et al., 2021; Herédia-Colaço and Rodrigues; 2021; Bonfanti et al., 2021; Hidalgo et al., 2022) hotels found that they could not rely on markets they previously depended on. Also, as experienced in other countries (Yacoub and El Hajjar, 2021) tourists preferred short-distance flights for their holidays. Consequently, there has been a significant decline in the number of tourists, and the hotels had to turn to closer markets (Belarus, Bulgaria, Germany, the UK, and the Ukraine), and domestic tourists, and for local customers for the daily use of hotel facilities to generate income for the hotel. Table 11.4 also showed that there was no major change in the hotel market in terms of business travelers. In parallel with the measures described by Gursoy et al. (2020), the use of technology enabled the hotels to increase customer satisfaction and trust.

TABLE 11.5 Marketing Strategies of Hotels After the COVID-19 Pandemic.

The main theme	Subthemes	Views of the participants (P)
Advertising and promotion	Digital marketing Social media advertisements	**P 5:** *We tended to place greater importance on Google advertisements. When customers carry out Google searches, we appear at the top. Besides, with the sponsorship advertisements, we happen to reach all the customers searching for the digital platforms.* **P 9:** *We emphasize hotel hygiene and safety with social media advertisements. We started using the social media accounts more actively after the COVID-19. We share all the promotion and campaign information through social media.*
Targeting different markets	Activities with the agencies Individual customers	**P 1:** *We enriched our customer profile by working with different agencies. We signed an agreement with Jet2, the second largest agency in the UK.* **P 10:** *We started Daily campaigns to attract individual customers to the hotel for the weekends. We started different campaigns with more reasonable food-beverage prices.*
The Extension of the Product Variety	MICE (Meeting, Incentive, Conference, and Exhibition) Market Spa-Wellness, Sports Facilities	**P 2:** *We established a good sales team for meetings, events, and weddings. We were preferred for the weddings especially during the summer. At the end of the season, several company meetings were held at our hotel.* **P 7:** *We attracted spa and wellness customers to the hotel for daily use. We arranged different campaigns for membership to the sports facilities, and the use of the beach. We organized special weight-losing campaigns with accommodation at discounted rates. As the health consciousness and the importance of sports have increased after the COVID-19 we turned these trends to our advantage.*

Based on Table 11.5 it can be seen that there have been significant changes in the sales and marketing strategies of the hotels. As in the case of hotel managers in other countries (e.g., Hao et al., 2020), hotel managers in Turkey have turned attention to the effective use of social media. As Ayyildiz (2020a, 2020b) found social media can be significantly effective in the marketing of tourism and hospitality establishments.

The hotels reviewed their suppliers (agencies) and made improvements in their customer profiles. They began using social media more actively to reach customers. They have made changes in their target markets, not only geographically, but also in terms of the nature of the market. As in the case of Spain (Hidalgo et al., 2022), the hotels have turned their faces more toward the local customers and the MICE market. To be able to cater to the MICE market, they reshaped their sales and marketing teams. They also opened up their hotels to the locals and domestic customers for the use of spas, wellness, and sports centers to cover some of the costs and enable the hotels to keep running with the additional income accruing. This appeared to be a sensible move as the health and sports consciousness of customers have increased after the COVID-19 Pandemic. Table 11.5 shows that the hotels have reviewed almost all of their marketing mix (7Ps) strategies (Koc and Ayyildiz, 2021b) and developed significantly different offerings.

Table 11.6 demonstrates that there have been significant changes in the competitive strategies of hotels as well. They have become more resilient and aggressive in their approach. They have become more competition oriented and kept an eye on competitors' level of prices and campaigns. This new more hands-on approach is in terms of close monitoring of the internal operations, the competitors, and the overall market, in general, is extremely needed for Turkish hotels. Due to the relatively high power distance in Turkey, the subordinates resort to indirect messages when they communicate failures and problems to their superiors (Koc, 2013b). This, in turn, results in delays in the solving of problems, causing customer dissatisfaction, the spread of negative WOM, and eventually leading to the loss of competitive advantage (Koc, 2102b).

Also, it needs to be noted that this new heavy reliance on price competition though may help certain hotels to survive in the short term, it may have significant implications in the long run.

However, as the hotels tend to resort to price discounts more heavily in the aftermath of the COVID-19, price-based campaigns may lead to price wars eventually and harm all the players in the market (King et al., 2022).

As the case of Chinese hotels (Hao et al., 2020) explained above, the hotels in Turkey concentrated on customer loyalty as well. This is a sensible thing to do, as it is generally accepted in marketing, trying to win new customers may be five times more expensive than serving the loyal customers (Peters, 1978), and increasing customer retention by 5% may result in increases in profits by 25% to 95% (Reichheld and Sasser, 1990). Unlike Peters (1978), Vandenbosch and Dawar (2002) calculated the cost of

TABLE 11.6 Competition Strategies of Hotels After the COVID-19 Pandemic.

The main theme	Subthemes	Views of the participants (P)
Price	Discounts	**P 3:** *We tried to offer advantages to customers by monitoring the prices of our competitors and offering discounted prices.*
		P 8: *We offered more reasonable prices to the agencies compared with the hotels in the same segment as us.*
Innovation	The Establishment of Call Centers	**P 1:** *We established our call center department as direct selling of the rooms without the intermediaries is more beneficial.*
	Changes in Room Design Online check-in Digital Menus	**P 5:** *For those customers who carried out their jobs online we placed desks and stationery in the rooms and tried to create a work atmosphere. The occupancy rate has increased with the customers who had their holidays while they were able to work.*
		P 9: *The membership program users of the hotel were enabled to make their check-ins contactless. Additionally, the use of QR-code contactless menus at the hotel bars and restaurants increased guest satisfaction.*
Campaigns	A-la-carte restaurants	**P 2:** *We increased the promotion of A-la-carte restaurants on social media. We offered packages for people who stay two nights, such as offering a free a-la-carte dinner.*
	Campaigns Directed at Business Traveler Guests Business Spa and Wellness Packages	**P4:** *We offered advantageous accommodation for business travelers through loyalty cards. They can benefit from the discounts where in the world they stay in a group's hotels. Additionally, with the bonuses, they collect they may have the opportunity to stay free of charge at our hotels.*
		P 7: *We designed more reasonable spa and wellness packages compared with our competitors. There was a high level of demand.*

acquiring new customers much higher between 30 and 40 times more. It may be also true that selling to an existing customer can up to 14 times higher than the probability of selling to a new customer (Farris et al., 2010).

11.5 CONCLUSIONS

The study demonstrated how managers developed strategies in a dynamic market during an on-going crisis. Hence, the findings of the study may be useful for preparing junior managers for prospective senior positions.

The study showed that the operational, marketing, and competition strategies of hotel chains operating in Turkey bear resemblances to strategies of hotels in various other countries. This may be due to the fact that the hotels studied in the study were all large chains and they may have been better at accumulating and disseminating information and knowledge globally. Future studies may investigate to see whether there are differences of competition and marketing strategies between hotel chains and individual hotels.

The study also found that the COVID-19 pandemic made the hotels more resilient and active. The hotels have paid more attention to their customers, competitors, and their own marketing mix offerings. It is believed that the COVID-19 pandemic has increased customer- and marketing orientations of hotels operating in Turkey. This may mean that even after the pandemic the hotels may be able to embrace challenges posed by the environment in a more efficient and effective manner.

Future studies may also investigate the hotels that have not been able to survive through the COVID-19 pandemic, in terms of the strategies they developed or did not develop. Also, as the study showed that the hotels developed reactive strategies toward the pandemic, it may be worthwhile to investigate whether this experience has increased hotels' proactive abilities.

Furthermore, the strategies show a wide spectrum of activities across several business functions. Future studies may investigate whether there is a relationship with the ability to cope and survive through the pandemic and the managers' abilities to work cross-functionally.

KEYWORDS

- **hotel chains**
- **COVID-19 pandemic**
- **marketing strategies**
- **competition strategies**
- **target markets**
- **marketing orientation**

REFERENCES

Ayyildiz, A. Y. Marketing Strategies of Hotels During the COVID-19 Epidemic: The Case of Kuşadası. *BMIJ* **2020a,** *8* (3), 3328–3358.

Ayyildiz, A. Y. The Using of Social Media in Hotel Businesses as a Marketing Tool: A Research on Afyonkarahisar Thermal Hotels. *Third Sector Soc. Econ. Rev.* **2020b,** *55* (3), 1486–1503.

Barber, B.; Odean, T. (2001). Boys Will Be Boys: Gender, Overconfidence, and Common Stock Investment. *Quart. J. Econ.* **2001,** *166*, 261–292.

Bonfanti, A.; Vigolo, V.; Yfantidou, G. The Impact of the COVID-19 Pandemic on Customer Experience Design: The Hotel Managers' Perspective. *Int. J. Hosp. Manage.* **2021,** *94*, 102871.

Bowen, W. M.; Bowen, C. C. Typologies, Indexing, Content Analysis, Meta-Analysis, and Scaling As Measurement Techniques. In *Handbook Of Research Methods in Public Administration*; Miller, G. J., Whicker, M. L., Eds.; NewYork: Marcel Dekker Inc, 2002; pp 51–86.

Farris, P. W.; Bendle, N.; Pfeifer, P. E.; Reibstein, D. *Marketing Metrics: The Definitive Guide to Measuring Marketing Performance*; Pearson Education, 2010.

Fletcher, J. Economics And Forecasting—Economic Impact. In *Tourism Marketing Management Handbook*; Witt, S. F., Moutinho, L., Eds.; Prentice Hall: Englewood Cliffs, 1995.

Gursoy, D.; Chi, C. G.; Chi, O. H. *COVID-19 Study 2 Report: Restaurant and Hotel Industry: Restaurant and Hotel Customers' Sentiment Analysis. Would They Come Back? If They Would, WHEN? (Report No. 2)*. Carson College of Business, Washington State University, 2020.

Han, H.; Lee, S.; Kim, J. J.; Ryu, H. B. Coronavirus Disease (COVID-19), Traveler Behaviors, and International Tourism Businesses: Impact of the Corporate Social Responsibility (CSR), Knowledge, Psychological Distress, Attitude, and Ascribed Responsibility. *Sustainability* **2020,** *12* (20), 8639.

Hao, F.; Xiao, Q.; Chon, K. COVID-19 and China's Hotel Industry: Impacts, A Disaster Management Framework, and Post-Pandemic Agenda. *Int. J. Hosp. Manage.* **2020,** *90*, 102636.

Herédia-Colaço, V.; Rodrigues, H. Hosting in Turbulent Times: Hoteliers' Perceptions and Strategies to Recover from the COVID-19 Pandemic. *Int. J. Hosp. Manage.* **2021,** *94* (102835), 1–12.

Hidalgo, A.; Martín-Barroso, D.; Nuñez-Serrano, J. A.; Turrión, J.; Velázquez, F. J. Does Hotel Management Matter to Overcoming the COVID-19 Crisis? The Spanish Case. *Tour. Manage.* **2022,** *88*, 104395.

Horwath HTL European Chains & Hotels Reports. 2019. https://horwathhtl.com.tr/en/publication/european-chains-hotels-report-2019/ (Accessed 25 Jan 2022).

Joy, L.; Carter, N. M.; Wagener, H. M.; Narayanan, S. The Bottom Line: Corporate Performance and Women's Representation on Boards, Catalyst Research Report, 15 Oct 2007. http://www.catalyst.org/knowledge/bottom-line-corporate-performance-and-womensrepresentation-boards

Karadeniz, E.; Beyaz, F. S.; Ünlübulduk, S. N.; Kayhan, E. Evaluation of the Impacts of COVID-19 Pandemic on Tourism Sector and the Applied Strategies: A Research on Hotel Managers. *Turk. J. Tour. Stud.* **2020,** *4* (4), 3116–3136.

Kenny, J.; Dutt, C. S. The Long-Term İmpacts of Hotel's Strategic Responses to COVID-19: The Case of Dubai. *Tour. Hosp. Res.* **2021,** 14673584211034525.

Kim, J. J.; Han, H. Saving The Hotel İndustry: Strategic Response to the COVID-19 Pandemic, Hotel Selection Analysis, and Customer Retention. *Int. J. Hosp. Manage.* **2022,** 103163.

King, C.; Wa I.; Julian C. Reimagining Resilience: COVID-19 and Marine Tourism in Indonesia. *Curr. Iss. Tour.* **2022,** 1–17.

Koc, E. A Review of Country Competitiveness, Tourism Industry Performance and Research Performance Relationships. *Competitiveness Rev.* **2009,** *19* (2), 119–133.

Koc, E. Inversionary and Liminoidal Consumption: Gluttony on Holidays and Obesity. *J. Travel Tour. Market.* **2013,** *30* (8), 825–838.

Koc, E. Do Women Make Better in Tourism and Hospitality? A Conceptual Review from a Customer Satisfaction and Service Quality Perspective. *J. Qual. Assurance Hosp. Tour.* **2020a,** *21* (4), 402–429.

Koc, E. *Cross-Cultural Aspects of Tourism and Hospitality: A Services Marketing And Management Perspective*; Routledge: Abingdon, 2020b.

Koc, E.; Ayyildiz, A. Y. Culture's Influence on the Design and Delivery of the Marketing Mix Elements in Tourism and Hospitality. *Sustainability* **2021a,** *13* (21), 11630.

Koc, E.; Ayyildiz, A. Y. An Overview of Tourism and Hospitality Scales: Discussion and Recommendations. *J. Hosp. Tour. Insights*. Article in Press, **2021b**. https://doi.org/10.1108/JHTI-06–2021–0147.

Lee, C. C.; Chen, M. P. The Impact of COVID-19 on the Travel and Leisure Industry Returns Some International Evidence. *Tour. Econ.* **2021,** 1354816620971981. DOI: 10.1177/13548166209719813

Lida, A. Gender Inequality in Japan: The Status of Women, and Their Promotion in the Workplace. *Corvinus J. Int. Aff.* **2018,** *3* (3), 43–52.

OECD Turkey–Tourism Economy, 2020. https://www.oecd-ilibrary.org/sites/f3b16239-en/index.html?itemId=/content/component/f3b16239-en#:~:text=Tourism%20is%20one%20of%20Turkey's,total%20service%20exports%20in%202018. (Accessed 31 Jan 2022).

Peters, T. *Thriving on Chaos*; Alfred A. Knopf: New York, 1987.

Reichheld, F. F.; Sasser, W. E. Zero Defeofions: Quoliiy Comes to Services. *Harv. Busi. Rev.* **1990,** *68* (5), 105–111.

Remington, J.; Kitterlin-Lynch, M. Still Pounding on the Glass Ceiling: A Study of Female Leaders in Hospitality, Travel, and Tourism Management. *J. Human Resour. Hosp. Tour.* **2018,** *17* (1), 22–37.

Sigala, M. A Bibliometric Review of Research on COVID-19 and Tourism: Reflections for Moving Forward. *Tour. Manage. Perspect.* **2021,** *40*, 100912.

Smart, K.; Ma, E.; Qu, H.; Ding, L. COVID-19 Impacts, Coping Strategies, and Management Reflection: A Lodging Industry Case. *Int. J. Hosp. Manage.* **2021,** *94*, 102859.

Spanaki, M. Z.; Papatheodorou, A.; Pappas, N. Tourism in the Post (?) COVID-19 Era: Evidence from the Hotel Sector in the North East of England. *Worldwide Hosp. Tour. Themes* **2021**.

Teikoku Databank COVID-19-Related Bankruptcies, 2022. https://www.nippon.com/en/japan-data/h01215/ (Accessed 25 Jan 2022).

Vandenbosch, M.; Dawar, N. Beyond Better Products: Capturing Value in Customer Interactions. *MIT Sloan Manage. Rev.* **2002,** *43* (4), 35–42.

Yacoub, L.; ElHajjar, S. How Do Hotels in Developing Countries Manage the Impact Of COVID-19? The Case of Lebanese Hotels. *Int. J. Contemp. Hosp. Manage*. **2021,** *33* (3), 929–948.

Yıldırım, A.; Şimşek, H. *Research Methods in Social Sciences*; Seckin Yayincilik: Ankara, 2013.

Yiwei, W.; Khakan, N.; Guilherme, F.; Osama, F. A. Influence Of COVID-19 Pandemic on the Tourism Sector: Evidence from China and United States Stocks. *Curr. Iss. Tour.* **2021,** 1–16.

CHAPTER 12

Market Orientation in the Relationship Between the COVID-19 Pandemic Process and City Marketing: An Evaluation in Terms of Crisis, Consumer Behaviors, and Spatial Planning

ABDULLAH ERAVCI

Boyabat Vocational Higher School, Member of Department Wholesale and Retail, Sinop University, Sinop, Turkey

ABSTRACT

City—it is a market place for its residents and visitors and as a business it has to meet the needs of its consumers. The city tends to grow as long as it meets the needs and expectations of its consumers. The COVID-19 pandemic process has made it difficult to meet consumer needs in the city and the flow of life. Consumers and city managers have increased their concerns over the market. The inadequacy of financial resources and the fact that the pandemic crisis turns into an economic crisis and affects life invites market orientation in city marketing. Market orientation is a consumer-based strategy for businesses to maintain the city's growth strategy in line with their growth goals. Market-oriented planning against crises in highly populated city life also seems to be a strategy to respond to the concerns of decision makers.

The purpose of your study is to raise awareness on market-oriented planning that cares about the consumer who is moving away from crowded living spaces due to the pandemic. As a method, literature studies were carried out. The information obtained is that the COVID-19 pandemic process is forcing

Dynamics of the Tourism Industry: Post-Pandemic and Post-Disaster Perspectives and Strategies.
Gül Erkol Bayram and Anukrati Sharma (Eds.)

consumers to change. It is recommended to take a market-oriented approach to upon the pandemic crisis, consumer behavior, and spatial planning in city marketing by keeping up with the change in the consumer. It is suggested to bring a market-oriented approach on the pandemic crisis, consumer behavior, and spatial planning in city marketing by keeping up with the change in the consumer.

12.1 INTRODUCTION

The COVID-19 pandemic period affects the country's economies in macro terms as well as the city economies in micro sense (Akbas, 2020). Cities are the driving force of the country's economies (Anholt, 2006). The local impacts of the pandemic process are deeply felt by the consumers in the cities. The tendency of consumers to move away from the city or to leave the city due to the pandemic is a very important problem in terms of the sustainability of city life. To contribute to the solution of this problem, the topic of market-oriented planning is discussed in this study. The goal is to generate benefits without driving the consumer away from the city. Creating value for consumer benefit through market-oriented strategies for the growth of cities and businesses in cities is in line with the crisis process (Amadasun and Mutezo, 2021). Creating value with market-oriented strategies for the benefit of consumers is appropriate in terms of response to the crisis process so that cities and businesses in cities can continue to grow. Since the object of the crisis is the consumer, social and political instability directly affects consumers' life.

It is clear that the COVID-19 pandemic process has a negative impact on city marketing. Throughout history, there has always been a desire by those living in pandemics to leave cities (Martinez and Short, 2021). However, the city is the most important part of human or consumer life. Two issues can be mentioned that can reduce the negative impact on city marketing during the pandemic process. The first is the crisis caused by the pandemic. The second is consumer behavior, which is associated with the first. The social and economic crisis in city marketing during the pandemic period has been an important problem for the consumer. This study has been carried out on the assumption that the right strategy to contribute positively to the solution of the problem was a market-oriented strategy.

Market focus: it is an approach that collects information by focusing on consumers and competitors and provides orientation by integrating the information it collects into the business (Hansen et al., 2006). At this point, the question of whether the city is a business comes to mind. The knowledge that city managers accept the city as a business, follow consumers and competitors, and produce market-oriented strategies is included in the literature (Box, 1999). According to this information, city managers are required to carry out appropriate improvement and development studies in the emerging crisis conditions. Adopting market-oriented strategies during times of economic crises is the best method in terms of performance (Sternad, 2012; Hansen et al., 2006). The city can be flexible with its residents and take measures to alleviate their distress (Martinez and Short, 2021). This information shows that a market-oriented approach can be taken in the context of crisis, consumer behavior, and spatial planning by associating city marketing with the COVID-19 pandemic process.

12.2 COVID-19 PANDEMIC PROCESS

The COVID-19 outbreak emerged in China in December 2019, with first-degree measures introduced in February 2020, resulting in the banning of collective programs, the closure of schools, and travel restrictions in many cities (Zheng et al., 2021). Within a month, these applications spread to almost all countries. It has had crippling effects for business, leading to the collapse of gigantic economies and rising unemployment levels (Opute et al., 2020).

As of the end of December 2021, the pandemic process and crisis has continued, yet. It is not possible to evaluate the process precisely because the results have not been evaluated, and to reach a conclusion-oriented judgment about the process (Wodak, 2021). According to the World Health Organization, there were 544,759 confirmed deaths as of December 29, 2021. The number of confirmed cases for the specified day is around 282,000 (WHO, 29 December, 2021). The number of cases and mortality rates are constantly changing. Countries are running a vaccination campaign to reduce cases and deaths.

Due to the fact that the pandemic process is not over, discussions continue about the social and economic dimensions of the pandemic crisis. At the heart of its social dimension is the consumer and the city. At the heart of its economic dimension are businesses. According to the report prepared by CRS (Congressional Research Service) and updated in November 2021, the global economic cost of COVID-19 exceeded $90 billion (CRS, 2021:1).

Under the social and economic costs of the pandemic, there are businesses, cities and consumers associated with both.

During the pandemic process, a kind of pandemic economy is applied. Businesses are under the influence of restrictions imposed due to the pandemic. While producers, employees, and consumers are affected by the process, the companies are at the center of them. From this point of view, there are those who say that the sector has become fragile and that the pandemic is heading toward the fragile sector (Karakurt et, al., 2015). Local governments, in other words, cities, feel the sector's problems most closely.

The consumer prefers the city to meet their needs as long as the conditions allow. According to Kavaratzis (2004), cities promise economic vitality and cultural services for human life. If these two features disappear, the city will lose its importance in the eyes of the target audiences and opportunities will become a threat. Opportunities and threats will come from outside according to the strengths and weaknesses of the city. In this case, it is worth examining the topic of city marketing to see what pandemic conditions have made changing in the city.

12.3 CITY MARKETING

City marketing has two dimensions: city and marketing. Although it is a geographical place at first glance, the city also has a functional feature (Kavaratzis, 2004). City with two features: It is a geographical and administrative place that has natural, historical, and cultural heritage on it and meets the needs of visitors, students, investors, new settlers, and other target audiences, especially its residents, with the goods and services it produces. In terms of meeting the needs of the target audiences, the city is a consumed product and market place. Marketing is an exchange process and requires a manufacturer/seller and buyer relationship. The seller and the buyer carry out the exchange by meeting or agreeing in a venue. This place is the city, and the shopping is marketing business.

City marketing entered the literature as a concept in 1969. Since the 1970s, the experience of the private sector in the marketing of goods and services has been utilized in city marketing. Since city marketing is a field of marketing science, it takes its basic knowledge from marketing science and has a rich and developable product range as a subdiscipline (Parker et al., 2015). The city is geography in a spatial sense. Kotler and his followers believe that "place" can be marketed in ground marketing (Parker et al.,

2015). The geography of the city has a wide variety of unique products that can be marketed to target audiences, and these products have the potential to be marketed as goods and services to target audiences.

Cities have the right to market the city's place quality as a product and to use its competitive advantage. They have cultural opportunities to market (Zhang and Zhao, 2009). They can make marketing planning on the appearance, historical, and cultural attractiveness of the city by making choices between functional and nonfunctional features.

This chapter on city marketing has so far examined the normal functionality of city marketing. During the pandemic process, local governments started to support the marketing efforts of the private sector in the city. In other words, the network of cooperation with the sensitivity and participation of public institutions and organizations in the city has expanded and the power of the state has been partially transferred to the city. With the years 2003 in Turkey, the transfer of authority from the center to the cities has gained momentum once again with the pandemic. Cities are not only responsible for zoning or urban planning in terms of local government, they also have to think about health. The pandemic process has created an important opportunity to rethink urban planning to ensure health and hygiene in the city or to create appropriate environmental conditions (Martinez and Short, 2021).

The shift in consumer behavior during the pandemic, the trend toward housing or housing, is reminiscent of the romantic American-style suburbs of the 1838s in urban design. The suburbs are located in the countryside away from the crowds of the city. Less socially restrictive, fun, healthy, summer, and villa-type houses are characteristic of the suburbs (Archer, 1983). Today, during the pandemic process, people tried to have detached houses with gardens, to make their own bread in their own homes, to produce as many products as possible, and to spend time safely around the house with their family members. These behaviors of people as consumers under pandemic pressure suggest that approaches in city marketing should also be revised.

12.4 MARKET ORIENTED IN PANDEMIC PROCESS

The concepts of market-oriented and marketing-oriented are concepts that have differences between them and do not mean the same. Marketing oriented and marketing function include focus on personnel and activities. Market focus is toward all employees, consumers, competitors, internal and external processes (Lettice et al., 2014). According to research conducted by Lettice et al., the concept of market focus dates back to the 1990s. The aim of market focus is to improve the performance of the enterprise.

According to the information provided by Lettice et al, it is seen that researchers focus on different issues such as consumer, decision processes, information gathering activities, business culture, operating capacity, strategies, and consumer behavior in market-oriented. It is possible to talk about a common market-oriented strategy that will be the answer to all of this. Different market-oriented conditions may bring different trends (Lettice et al., 2014). The most important advantage of being market-oriented is that it keeps the business open to innovations, consumer expectations, from the delivery of goods and services to consumer satisfaction (Na et al., 2019). The consumer orientation requires the business to be proactive in all its marketing activities.

The COVID-19 pandemic has brought with it some difficulties. Those difficulties reminded businesses of the need to turn to and adopt proact and innovative marketing strategies in overcoming challenges (Opute et al., 2020). COVID-19, or coronavirus expressed in its broadly worded year, has caused a worldwide crisis, as stated by Opute et al.

12.4.1 MARKET-ORIENTED STRATEGY IN CRISIS

The crisis is a low probability and highly effective situation that is perceived by stakeholders as a threat to the life of the enterprise. Crises include corporate collapses, government rescue operations, fluctuations in housing and foreign exchange markets, layoffs, and inventory problems (Lettice et al., 2014).

Although the framework of the crisis caused by the COVID-19 pandemic process is not yet fully predicted, there are crisis situations. Due to the fact that the crisis has not been concluded, it is too early for all different rhetoric to be discussed and a full assessment can be made (Wodak, 2021). There are also differences in approach between countries, regions and cities, and product and market-oriented changes in the current crisis, and ongoing concerns about both production and market sustainability (United Nations, 2020).

Despite the uncertainties and different practices related to the crisis, various suggestions are made in academic research on the issue of being transparent to consumers, offering options such as suitable, inexpensive products, and safe living spaces (ECLAC, 2020 Covid Report). As a result of a study conducted in the days of COVID-19, it is proposed to increase the variety of accommodations during the pandemic process and to offer options to visitors (Jiang and Wen, 2020). The consumer generally wishes to meet their needs without contact and to be environmentally protected.

12.4.2 MARKET-ORIENTED ON CONSUMER BEHAVIORS

Market focus during the pandemic process requires monitoring consumer behavior. Faced with the fear of transmission and death, the consumer may exhibit irrational behavior and group behavior (Song et al., 2020). According to the research conducted by Song et al., factors such as allocation of resources, misuse, changing prices, panic status, and incompatibilities in the information provided are factors that affect the consumer.

A number of empirical studies have been conducted to understand and analyze consumer behavior, dating back to the Russian scientist Pavlov. These studies were applied to consumer behavior by Kotler and explained with rational psychology. Accordingly, the consumer is motivated, signs are followed, consumer reaction is received, and reinforcements are made (Kotler, 1965). The work carried out by Kotler was later developed and in 1974 the three-stage stimulus-organism and response model developed by Mehrabian and Russell was born. In a study on consumer behavior during the COVID-19 pandemic in 2020, the model was tested and the consumer's self-isolation intention and unusual purchasing behavior were observed (Laato et al., 2020).

In relation to city marketing, what the target audiences are turning to during the pandemic process, what they prefer in terms of food and drink, what choices they make in terms of housing or accommodation, and what methods they use are the subject of research. In a study conducted by Zaghol and Pensioner (2020), the trend of searching for housing in rural areas under the influence of the COVID-19 pandemic approached 70% compared to urban centers. In other words, two out of three people who wanted to buy housing sought housing in rural areas far from the city's crowds. In this sense, it has become a situation where it is desirable to spend time in the village for those who have a village. The preferences of living spaces such as housing and cities described above on consumer behavior indicate the importance of spatial planning.

12.4.3 MARKET-ORIENTED IN SPATIAL PLANNING

One of the most important lessons that can be learned during the COVID-19 pandemic is the design and planning of the city. When the researches in the first stage of the pandemic process are examined, four issues are noted: environmental quality, socioeconomic effects of the pandemic, governance, transportation and city design (Sharifi and Khavarian-Garmsir, 2020; Pantic et al., 2021). At the center of socioeconomic effects is the person who consumes. Those who produce and trade the product will be market-oriented, considering the consumer. When management or governance is market-oriented and transportation is market-oriented, the planning of the city has to be market-oriented in a spatial sense.

It is stated that spatial planning in settlements was born in the 18th century in scientific terms, gained momentum with the laws passed starting from developed countries in the 20th century, and in some countries legal preparations began due to the pandemic (Pantic et al., 2021). Due to infectious diseases such as malaria and cholera, modern planning reportedly began in cities in the 19th century (Klaus, 2021). Spatially, settlements have to make extensive, comprehensive, and sustainable plans (Kırgız, 2017). The adoption of laws on city planning is an indication of sustainability intent. Sustainable city planning refers to the planning of infrastructure and superstructures (Kavaratzis, 2004).

Jones and Iverson (2012) state that one of the measures to be taken against pandemics is environmental policy measures. What environmental policies can be in terms of planning has emerged during the pandemic process. People who move away from the streets and crowded places, produce and consume are confined to their homes, and the possible works are carried out remotely or from homes. With the relocation of the work environment to the house, family members become a kind of partner in the work done. In this case, common relationships increase in the family and the demands for activities together in the family's own private environment increase. In a way, previous work habits are changing and social relations are taking on another dimension and environments are being sought that are more suitable for the new situation.

Research into the impact of the pandemic on consumer behavior indicates that people are spending more time at home and engaging in "home improvement" efforts. The concept of home improvement is used in terms of the development of home-related hobbies (Tekin, 2020). According to Tekin (2020), outside spending decreases during the pandemic process and the advantage of this decrease is investment in the form of maintenance and repair in the home. People show walking and cycling behavior in the immediate vicinity with their focus on the home and family environment. After the pandemic, more walkable and more suitable city designs for cycling may come. According to observations, people's tendencies in this direction have strengthened (Martinez and Short, 2021).

A study in China highlights continued growth and the integration of markets for rapid growth (Zhang and Zhao, 2009). Since people are forced to live an isolated life and apply the distance rules due to the pandemic, the growth of the city will be possible by producing housing and accommodation opportunities that are suitable for the conditions. The centralization of housing and other living spaces in the city makes it difficult to prevent close contact in pandemics and outbreaks (Daneshpour, 2020). According to Daneshpour, the solution to preventing close contact is to plan the need for housing at all levels and income levels.

12.5 CONCLUSIONS

In this study, the city marketing and crisis in the axis of marketing, changing consumer behaviors, and the need for spatial planning were investigated during the COVID-19 pandemic process. The aim is to emphasize the consumer's need for safe living spaces against regional or global epidemics and to contribute to marketing science. It is important to develop a market-oriented entrepreneurial city approach as a solution to pandemics and similar crises under the pressure of COVID-19.

By replanning the city, especially focusing on housing and accommodation sales, accompanied by information compiled from the literature, it can be effective for city marketing. Since location and housing quality, price reductions, increase in housing and accommodation variety, and consumer behavior are interrelated, it is useful to evaluate consumer behavior in terms of rationality and emotionality and to offer options according to the expectations of the target audience.

As a result of the research, the following suggestions were made:

- The COVID-19 pandemic process reveals the need for the city's life to be reshaped.
- The intertwining of shopping centers, housing, and other living spaces poses safety and health problems.

- Diluted structures are needed in cities, including natural disasters that will create crises, and safe neighborhoods and environments with completed infrastructure and superstructure in terms of transportation and access.
- It is assessed that housing and market-oriented city marketing is needed to ensure the well-being and life of the city's inhabitants safely.
- Market-oriented approaches are proposed in terms of analyzing the impact of COVID-19 on consumers and markets, discussing the results on large platforms, urban marketing and the sustainability of city life by bringing together different disciplines and stakeholders.

KEYWORDS

- **city marketing**
- **consumer**
- **city planning**
- **COVID-19**

REFERENCES

Akbas, E. Kovid-19 ve Sonrası: Disiplinlerarası Bir Yaklaşım: COVID-19'un Toplumsal Etkileri, *Ankara Yıldırım Beyazıt Üniversitesi Uluslararası İlişkiler ve Stratejik Araştırmalar* (ULİSA) Enstitüsü **2020,** *12* (3), 1–68.

Amadasun, D. O. E.; Mutezo, A. T. Effect of Market-Driven Strategies on the Competitive Growth of SMEs in Lesotho. *Res. Square* **2021,** 1–20. DOI: 10.21203/rs.3.rs-439382/v1

Anholt, S. How the World Sees the World's Cities. *Place Branding* **2006,** *2* (1), 18–31.

Archer, J. Country and City in the American Romantic Suburb. *J. Soc. Archit. Historians* **1983,** *42* (2), 139–156.

Box, R. C. Running Government Like a Business: Implications for Public Administration Theory and Practice. *Am. Rev. Public Admin.* **1999,** *29* (1), 19–43.

CRS – Congressional Research Service. *Global Economic Effects of COVID-19*, 2021. https://sgp.fas.org/crs/row/R46270.pdf (Accessed 1 Oct 2022).

Daneshpour, Z. A. *Out of the Coronavirus Crisis, a New Kind of Urban Planning Must be Born, Post Pandemic Urban and Regional Planning and the Lessons That Can Be Learned from Coronavirus Pandemic*, 2020. https://www.researchgate.net/publication/340491887 (Accessed 6 Oct 2021).

ECLAC—Economic Commission for Latin America and the Caribbean. United Nations, Special Report COVID-19, No: 4, 2020. https://repositorio.cepal.org/bitstream/handle/11362/45736/5/S2000437_en.pdf (Accessed 3 Oct 2022).

Hansen, E.; Dibrell, C.; Down, J. Market Orientation, Strategy, and Performance in the Primary Forest Industry. *Forest Sci.* **2006,** *52* (3), 209–220.

WHO-World Health Organization. *Coronavirus Disease (COVID-19) Pandemic*, 2021. https://www.who.int/emergencies/diseases/novel-coronavirus-2019 (Accessed 6 Oct 2021).

Jiang, Y.; Wen, J. Effects of COVID-19 on Hotel Marketing and Management: A Perspective Article. *Int. J. Contemp. Hosp. Manage.* **2020,** *32* (8), 1–16.

Jones, S. C.; Iverson, D. Pandemic Influenza: A Global Challenge for Social Marketing. *Health* **2012,** *4*, 955–962.

Karakurt, B.; Senturk, S. H.; Ela, M. Makroekonomik Kırılganlık: Türkiye ve Şangay Beşlisi Karşılaştırması. *Yönetim ve Ekonomi Dergisi* **2015,** *13* (1), 283–307.

Kavaratzis, M. From City Marketing to City Branding, *Place Brand.* **2004,** *1* (1), 58–73.

Kırgız, A. C. *Pazarlama Estetiği: Marka Şehir*; Türkmen Kitabevi: İstanbul, 2017.

Klaus, I. *Pandemics Are Also an Urban Planning Problem*, 2021. https://www.bloomberg.com/news/articles/2020–03–06/how-the-coronavirus-could-change-city-planning (Accessed 30 Dec 2021).

Kotler, P. Behavioral Models for Analyzing Buyers. *J. Market.* **1965,** *29* (4), 37–45.

Laato, S.; Islam, N.; Farooq, A.; Dhir, A. Unusual Purchasing Behavior During the Early Stages of the COVID-19 Pandemic: The Stimulus-Organism-Response Approach. *Journal of Retailing and Consumer Services*, **2020,** 57, 10**2224,**1–12.

Lettice, F.; Tschida, M.; Frontbencher, I. Managing in an Economic Crisis: The Role of Market Orientation in an International Law Firm. *J. Busi. Res.* **2014,** *67*, 2693–2700.

Martinez, L.; Short, J. R. The Pandemic City: Urban Issues in the Time of COVID-19. *Sustainability* **2021,** *13* (3295), 1–10.

Na, Y. K., Kang, S.; Jeong, H. Y. The Effect of Market Orientation on Performance of Sharing Economy Business: Focusing on Marketing Innovation and Sustainable Competitive Advantage. *Sustainability* **2019,** *11* (729), 1–19.

Opute, A. P.; Iwu, C. G.; Adeola, O.; Mugobo, V. V.; Okeke-Uzodike, O. E.; Fabola, O.; Jaiyeoba, O. The COVID-19-Pandemic and Implications for Businesses: Innovative Retail Marketing Viewpoint. *Retail Market. Rev.* **2020,** *16* (3), 83–98.

Pantic, M.; Cilliers, J.; Cimadomo, G.; Montano, F.; Olufemi, O.; Mallma, S. T.; Berg, J. Challenges and Opportunities for Public Participation in Urban and Regional Planning during the COVID-19 Pandemic—Lessons Learned for the Future. *Land* **2021,** *10* (1379), 1–20.

Parker, C.; Roper, S.; Medway, D. Back to Basics in the Marketing of Place: The Impact of Litter Upon Place Attitudes. *J. Market. Manage.* **2015,** *31* (9–10), 1090–1112.

Sharifi, A.; Khavarian-Garmsir, A. R. The COVID-19 Pandemic: Impacts on Cities and Major Lessons for Urban Planning, Design, and Management. *Sci. Total Environ.* **2020,** *749* (421391), 1–14.

Song, W.; Jin, X., Gao, J.; Zhao, T. Will Buying Follow Others Ease Their Threat of Death? An Analysis of Consumer Data during the Period of COVID-19 in China. *Int. J. Environ. Res. Public Health* **2020,** *17* (3215), 1–15.

Sternad, D. Adaptive Strategies in Response to the Economic Crisis: A Cross-Cultural Study in Austria and Slovenia. *Manag. Global Transitions* **2012,** *10* (3), 257–282.

Tekin, I. Ç. Pandemi Sürecinde Değişen Tüketici Davranışları. *Busi. Manage. Stud. Int. J.* **2020,** *8* (2), 2331–2347.

United Nations. *Impact of the COVID-19 Pandemic on Trade and Development: Transitioning to a New Normal*. 2020 United Nations Conference on Trade and Development, 2020.

https://unctad.org/system/files/official-document/osg2020d1_en.pdf (Accessed 1 Dec 2020).

WHO-World Health Organization. *Coronavirus Disease (COVID-19)* Pandemic, 2019. https://www.who.int/emergencies/diseases/novel-coronavirus-2019 (Accessed 1 Dec 2020).

Wodak, R. Crisis Communication and Crisis Management during COVID-191. *Global Discourse* **2021,** 00 (00), 1–20.

Zhang, L.; Zhao, S. X. City Branding and the Olympic Effect: A Case Study of Beijing, *Cities* **2009,** *26*, 245–254.

CHAPTER 13

Financing of Sports Tourism Marketing in India: An Approach Toward Sustainable Development

PARAG SHUKLA[1] and SOFIA DEVI SHAMURAILATPAM[2]

[1]*Department of Commerce and Business Management, The Maharaja Sayajirao University of Baroda, Vadodara, Gujarat, India*

[2]*Department of Banking and Insurance, The Maharaja Sayajirao University of Baroda, Vadodara, Gujarat, India*

ABSTRACT

Research in the field of sports tourism marketing for sustainable development is in a nascent stage, and few extant literatures are available while research studies are conducted taking into account the regional and global perspectives. Marketing of destination tourism with reference to the leagues/tournaments, infrastructure facilities, availability of equipments, complex, capacity, training institutes for athletes, technological advancements, and the role of governing bodies are the major aspects needed to promote and enhance sports tourism. Identifying and promoting a country or state as a hub of sports destinations involve huge investment and financing mechanisms given the large number of stakeholders under different hierarchy of authorities at local, national, and global contexts. Conventionally, theories in finance have validated the role of finance in economic development either through demand-following hypotheses or supply-leading hypotheses or a combination of both. This study attempts to find out the role of financing by different stakeholders in the expansion of sports as investment in human capital and how marketing can accomplish the goals as a driver of destination tourism

Dynamics of the Tourism Industry: Post-Pandemic and Post-Disaster Perspectives and Strategies.
Gül Erkol Bayram and Anukrati Sharma (Eds.)

of a particular region/state/country keeping into account different attributes of sports activities. The vital role sports sector plays in bringing sustainable economic development in terms of building human capitals, curbing crimes, revenue earnings, employability, and contribution to GDP of the nation will be taken into this research study.

13.1 INTRODUCTION

Tourism has a great potential to accelerate progress across the sustainable development goals with its significant impact on efforts to achieve other economic goals in terms of eradication of poverty, advancing gender equality, and protecting the environment (UNDP, 2018). Sports as a destination tourism is a new concept which envisages attracting tourist footfalls across the globe through multifaceted sports events, given the cultural and social milieu. Marketing of destination tourism with reference to the leagues/tournaments, infrastructure facilities, availability of equipments, complex, capacity, training institutes for athletes, technological advancements, and the role of governing bodies are the major aspects to promote and enhance sports tourism. Identifying and promoting a country or state as a hub of sports destinations involve huge investment and financing mechanisms given the large number of stakeholders under different hierarchy of authorities at local, national, and global contexts. As per the UNTWO Report (2019), sports tourism as a segment of tourism can help achieve sustainable development in a distinct way stemming from its characteristics in terms of engagement in physical activities, opportunities for interactions and scope for development potential in region-specific ranging from small events or light activities to major sports events, such as Olympic games, football, and rugby championships. Sports tourism may attract tourists as participants or audience, and in this regard, the destinations have significant role to add local values to them through the provision of best local experiences with successful leveraging of branding of the destination, infrastructures, and socioeconomic benefits. The development of sports tourism requires coordination and cooperation of all the stakeholders in the sport and tourism sectors, the role of the government, the private institutions as well as the involvement of communities in order to benefit the destination/region. The involvement of locals is crucial in enhancing tourism development through the hiring of locals, promoting their entrepreneurship, and utilizing resources which will be a source of income opportunities for them. In this connection, local people should be given proper training and skills to handle and organize sports tourism. To implement and materialize it, locals need proper finance and funding from banks and financial institutions in the form of micro, small, and medium enterprises through the direct and indirect involvement of local authorities and the government at large.

Finance has a significant role to play to meet the goals of sustainability—either pushed from the demand-side factors or driven by supply-side impetus or both. As we all know that the COVID-19 pandemic has hit all sectors and tourism sector is no exception. According to the latest data of UNTWO, it is found that global FDI into tourism has slashed down by 73.2% in the first half of 2020, as compared with its previous period, leading to a decline in the number of international tourists by more than 50% (UNTWO, 2020). It is also clear that sustainable tourism requires sustainable investment and is not confined to traditional investments to promote economic growth and development; but also investment in nontraditional dimensions, such as to enhance innovation, creation, and dissemination of different sectors of the economy.

This chapter delves into the key aspects of sports tourism and its integral role in sustainable development. This chapter has six different sections. Section 13.2 discusses a brief on the role of sports tourism as a driver of economic growth. Section 13.3 reports the review of literature on certain dimensions and issues of sports tourism in emerging and developed economies of the world. Section 13.4 explains the rationale of the study with major objectives highlighted. Section 13.5 discusses the policy imperatives and some recommendations. The last section concludes the chapter.

13.2 SPORTS TOURISM AS A DRIVER OF ECONOMIC GROWTH

Sports tourism possesses a long history, and it is rather symbolic that the century which witnessed the most substantial growth and development of sports tourism was heralded by the revival of the modern Olympic Games in Athens in 1896, and even today that has clearly reflected the importance and significance of contemporary sports tourism, involving many of the influences and characteristics (Weed and Bull, 2004). Generally, sports tourism is an integration of social, cultural, and economic processes with the mere involvement in several activities by different stakeholders in a given time and space. Sports tourism in its many different forms involves comprehensive activity and is one of the significant markets in the tourism sector at present.

We can say that the sports and tourism sectors have mutual advantages in linking sports and tourism which produces significant outcomes in terms of opportunities for locals and their involvement and extending tourism with recreation activities. In other words, the promotion and upliftment of sports tourism are associated with job creation, opportunities for the locals, expansion of infrastructures, preservation of the cultural heritage, social and traditional values, livelihood sources, utilization of resources, curbing the environment with the core concept of sustainability dimensions in social, economic, and environmental aspects thereby contributing the real and actual values to the economy. However, for successful sports tourism, both government and private players should evolve a sports marketing campaign by involving local communities, thereby promoting rare and local sports creating livelihood and economic empowerment in specific region.

In the context of global integration of economies, tourism, specifically sports tourism is a significant element for developing the economies of advanced and developing economies. It not only substantially aids the economy in terms of multiple economic values, cultural preservation, image, regional identification, brand values, and utilization of natural resources for leisure and businesses but also acts as a source of creation of jobs, regional development, acceleration of economic activity, and contribution to the GDP.

TABLE 13.1 Top 10 Tourism Destinations, Tourist Arrivals, and % Share of GDP.

Rank (2018)	Countries	Arrivals (millions)	Tourism direct GDP as a proportion of total GDP (%)	
			2010	2017
1	France	89.4	7.3	7.2
2	Spain	82.8	10.3	11.8
3	United States	79.6	2.7	3.0
4	China	62.9	4.3	4.5
5	Italy	62.1	5.7	6.0
6	Turkey	45.8	-	-
7	Mexico	41.4	8.4	8.7
8	Germany	38.9	4.2	3.9*
9	Thailand	38.3	3.2	7.0
10	United Kingdom	36.3	3.4	3.3

Source: Adapted from 1. (UNTWO): Tourism Statistics Data.
2. (UNTWO): World Tourism Network Organization.
(Note:*, relates to the data of 2015.)

Table 13.1 presents an overview of tourism destinations across the globe and the number of tourists' arrivals as per the data given by UNTWO Report 2020. Thus, it is very clear that the number of tourist arrivals is more in advanced countries, and some of the developing and emerging economies, namely, China, Turkey, and Thailand also have a substantial number of tourist arrivals. In terms of the contribution of tourism GDP as a proportion to the total GDP, Spain has a percentage share of 11.8% though stood second in the ranking of tourism destinations in 2020. Mexico and France also have fair percentages of 8.7 and 7.2%, respectively, in terms of the tourism direct GDP as a proportion of total GDP. We can say that tourist destinations are concentrated in advanced countries and may be associated with better infrastructure and easy access to facilities for tourists on the one hand, and also attraction from various destination spots on the other hand. Developing countries that have not come across in the top 10 destinations do possess attractive tourist spots and sources for sports tourism, which include mountaineering, rock climbing, rafting, and tracing of rivers among others to mention, offsetting the seasonality of tourism; however, as these economies have less developed infrastructures, lack of facilities and easy access to destinations, communication, they are delisted among top destinations. In this regard, promotion through funding, sponsorship, and schemes from institutional sources—government and state authorities, private and public, partnership firms, and organizations dealing directly/indirectly with tourism sectors are required at timely intervals to meet the changing economic, social, and political environment. Therefore, the very core concept of this volume is that finance plays a significant role in the promotion of the destinations that are endowed in different geographical areas and topography to reap the benefits to stakeholders in the host countries as well as to the tourists in particular.

13.3 REVIEW OF LITERATURE

Research in the field of sports tourism marketing for sustainable development is in a nascent stage and few extant literatures are available while research studies are conducted taking into account the regional and global dimensions. Sports tourism provides tourists with extraordinary active participation as competitors in the event or experience as a spectator, and irrespective of their involvement, sports tourism is all about the interaction of activities, people, and places (Weed and Bull, 2004). Sports tourism is a low-cost leisure activity, with significant benefits to health—physically and mentally, and has gradually gained prominence in countries in creating many sports activities in destination spots, such as cycling, mountaineering, rafting, even tracing of river, and rock climbing among others (Pouder et al., 2018; Cho et al., 2019; Yang et. al., 2020). Different stakeholders are involved in the successful functioning of sports as destination tourism. The successful organization of any events in sports tourism in a destination relates to the involvement of the locals and authorities. Kim and Kaplanidou (2019) using a sample of 301 Korean residents examined how the involvement of locals and residents helps in the functioning of mega events in a destination. They reveal that involvement in sports can leverage support for a mega sports event through the creation of positive perceptions of the impact of the event, thereby increasing the quality of life gained from that event. The success of mega sports events such as the Olympic Games depends upon the involvement of the local community with recognizable benefits and significant impact on the sustainable development of the host country in terms of social, economic, and cultural dimensions in particular (Kaplanidou and Karadakis 2010; Waitt, 2003; Gursoy and Kendall, 2006; Funk and James, 2001).

Global integration of economies also rose in sports culture across the globe which are sources to attract tourists in destinations governed by different geographical features and typography on the one hand and also significant economic growth in its areas, expansion in social and economic infrastructures, uplift of industries associated with sports supplies, opportunities to locals, resource allocation, heritage values and also overcoming the issues of seasonality in tourism sector at large (Bourdeau, et al., 2002; Taleghani and Ghafary, 2014; Gratton et. al, 2000; Wilson, 2006; Hinch, 2001; and Higham, 2005). Yang et al. (2020) in their study assess the sustainability of sports tourism with reference to Taiwan have highlighted four different dimensions of sustainability—economic sustainability, environmental sustainability, sociocultural sustainability, and institutional sustainability. The models found a feasible solution in the optimization of events in destinations with the imperative that promotion of sustainable sports tourism must be supported by governments and be protected by relevant laws and regulations. In a similar kind of research, sustainability of golf as sports tourism relates to five dimensions of its impacts, such as—environmental impacts, environmental management, environmental attitudes and behaviors, conflict of interests, and sustainable management and planning. Among these, environmental impact is found to be the most prolific content, with environmental attitudes and behaviors being the least frequent, though it is the most discussed topic in recent trends (Lopez-Bonilla, 2020). How sports tourism is significant to sustainability in terms of economic, social, and environmental dimensions is a big concern. It is because not all tourism's effects are positive and the excessive growth of sports tourism and its lack of planning over decades lead to some tourist destinations victims of its success and are testing their carrying capacity, which may lead to overtourism, degradation in environmental values, threatening to cultural heritage, downfall in socioeconomic status among others to mention (Seraphin et al., 2019; Shukla and Shamurailatpam, 2021). In other words, the rapid growth of the tourism sector, particularly in developing countries, opens new threats to the environment in the form of energy consumption, access to and utilization of natural resources, water, and increasing the quantum of waste which are some of the external externalities impacted through the expansion of the sector. Hence the concept of sustainability should be brought whenever any branches of the tourism sector are considered so that its impact can be mitigated or kept at a bare minimum.

Digitization makes it easier to access tourism services in a more comfortable and convenient way. Countries have introduced strategies to promote destinations/regions with the help of the Internet on the platform of digital marketing campaigns. The role of technology and social media specifically can be leveraged out to lesser known sports on those destinations' less tourist attractions. Digital marketing makes it possible to spread out networks and services at end destinations through various instruments, such as social media, blogs, branding, content marketing, creation of Apps, and also video marketing among others as marketing tools. Many government schemes and marketing campaigns have been developed from time to time to attract more tourists to destinations; however, there is a dire need to integrate it with heritage value in terms of traditional sports for a better inflow of tourists.

Finance is significant for development. There are established theories and schools of thought on finance that lead to development—either from the demand-following factors or supply-leading factors to push for. The involvement of all the stakeholders including the locals and governing authorities in particular are pertinent in making optimal allocation of the scarce resources for the development of sports tourism in destinations. To put here, the demand side is exclusively determined by the destination marketing of sports tourism through various digital platforms and country-specific features predefined by its values attached to topography or geographical areas attracting tourists; the supply side in general relates to the involvement of all the stakeholders to uplift the destinations/events with funding aid, investment, strategic management, and allocation of resources to enhance the inherent qualities of the destinations. In all, both the demand side and supply side factors are coherent and co-exist for optimal and sustainable sports tourism in a destination. The very concept of bringing finance is that with proper planning and predefined strategies, all the targets will not be successful unless proper funding and financing are there to fulfill all the objectives attached in any given destination for the promotion of tourism sector and sustainable economic development.

13.4 RATIONALE OF THE STUDY

Conventionally, theories in finance have validated the role of finance in economic development either through demand-following hypotheses or supply-leading hypotheses, or a combination of both. This study attempts to explore the role of financing by different stakeholders in the expansion of sports as an investment in human capital and how marketing can accomplish the goals as a driver of destination tourism of a particular region/state/country keeping into account different attributes of sports activities. The vital role sports sector plays in bringing sustainable economic development in terms of building human capitals, curbing crimes, revenue earnings, employability, and contribution to the GDP of the nation will be taken into this research study.

Given the fundamental framework, the chapter attempts to highlight the role of finance in the successful organization and implementation of sports tourism, which is one of the challenges toward sustainability goals. These objectives include (a) what is the nature of association between the role of financing, marketing of sports tourism, and its impact on sustainable economic development; (b) what are interlinkages between sports financing, destination marketing, and sustainability; (c) what are the various determinants of sports financing for sustainable development; and (d) what are the various recommendations for developing a framework of sports financing which an ecosystem harnesses sustainability. Based on these objectives, the following hypotheses are formulated.

Hypothesis 1: Financing for sports tourism will have a significant impact on sustainable economic development.

Hypothesis 2: There is a positive association between financing for sports tourism and sustainability.

Hypothesis 3: There is a positive association between the determinants of financing sports tourism and sustainable development.

Hypothesis 4: Technology and digitization of marketing have a significant impact on sports tourism.

Hypothesis 5: Sports tourism will overcome the issues of tourism seasonality.

Hypothesis 6: Destination marketing communication has a positive impact on sustainable development.

Hypothesis 7: Tourism beyond its carrying capacity may impact negative externalities such as overtourism or degradation of destination.

These hypotheses together framed the major broad area of research spelt out in this chapter. The chapter helps to understand how destination tourism with sports tourism as one of the dimensions can have sustainable development, meeting the objectives of predefined SDGs of UNDP provided proper source and way of financing for sports tourism is accommodated in each destination. To validate and generalize the hypotheses for policy framework, theoretical models should suit the combined efforts and the role of government, policy authorities, private firms, a partnership of public and private bodies, institutional arrangements in the form of participatory models, such as revenue sharing models, economic growth models, regional development models, macroeconomic models with employment generation, and welfare models among others. With these objectives, the outcome of the expansion of sports tourism can be reaped by all the stakeholders involved in the successful organization and linking the tourism–sports association in a more fruitful way. The very basic idea is that tourism, specifically finance plays a very significant role in linking sports and tourism in destinations, bounced back by the shortage or lack of infrastructures and facilities for attracting more tourists out of seasonality in endowed destinations, particularly in developing and emerging economies.

13.5 KEY DISCUSSIONS AND POLICY IMPERATIVES

This chapter serves as a framework of how the financing of sports tourism in destination and marketing of sports tourism can bring sustainable development. The following key points are highlighted as policy imperatives so far as sports tourism and its inherent role in bringing sustainable development to destinations are concerned.

- Integration of the economies across the globe through sports tourism activities can nurture the continued participation of the community at large and the habit of sports.
- In order to provide high-quality tourism services, the role of government is important in terms of encouraging the sports to citizens and also the tourism industry in qualitative services.
- Governments, private institutions, and organizations should provide timely resources in promoting, branding, and image of destinations through various schemes, sponsorship, and funding agencies that directly or indirectly push up the sports tourism in destinations.
- Marketing through the use of by-products of technology and digitization is significant in reaching out and building off-seasonality of the destinations through the use of social media, Apps, sites for tourism, and announcement in networking platforms.
- Promotion of sports tourism not only builds branding, image of destinations but also encourages sports activities to tourists as recreational events.
- Sports tourism has a direct contribution to the economic growth of the economies through the involvement of locals and community at large, the creation of jobs, employment generation, regional development, and contribution to GDP.
- Sports tourism has a substantial impact on meeting the sustainable development goals of the nation, thereby taking into account the responsibility, accountability, and appraisal of the negative externalities associated with overtourism and loopholes in unsuccessful planning in the implementation of sports tourism in the destination.
- There is a need to formulate and institutionalize a Sports Tourism policy as a National Agenda.
- The inflows of foreign direct investment in sports tourism should be enhanced for a larger coverage and extension of sports tourism at regional levels.
- Leveraging rare and ancient sports by targeting communities and locals as the main drivers to promote.
- A special budget allocation as a priority area is missing in the context of financing for sports tourism is concerned particularly in developing and emerging economies despite these economies possessing numerous scopes for expansion of sports tourism.

The above listed points are some of the major key areas of discussion and some policy imperatives to be tackled by authorities in due course of time with the target of sustainable development goals by nations. Considering the loopholes associated with unsuccessful planning in extending sports tourism at the regional level, one of the major hindrances is the lack of sources of funding agencies/specific funds or sponsorship meant to promote and upgrade sports tourism in regional/local areas. Therefore, this chapter tries to bring out the role of financing in promoting sports tourism as an important variant of tourism in which it not only increases the international movements of tourists in destinations but also offshoots the seasonality characteristics of tourism sector across destinations in the world, maybe either through demand leading or supply following impetus to bring forth. We can say that the demand-leading factors such as demand for the products and services of sports tourism are relatively less influential in developing countries, for example, India, wherein the supply-side factors such as funding from the state and local governments, institutions have to play a significant role to promote the sports tourism as an industry.

13.6 CONCLUSIONS

As discussed in this chapter, sports tourism is in its nascent stage and no such extant literatures were established, though the concept of sports tourism dates back to the Olympics in Athens in the 17th century. This chapter highlights the significance of sports tourism in meeting the sustainable development goals of nations as defined by the UNDP. Sports tourism has multifaceted benefits to the economies—economic, sociocultural, and environmental aspects bring them together. From the economic point of view, it aids job creation, employment opportunities, using skills of locals, and source of livelihood particularly to the community or people in the region/destinations. It also features to restore the cultural and heritage values of destinations through the organization of events associated with each cross-cultural geographical topography. Sports tourism nullifies the core characteristic of seasonality of tourism by organizing sports as the mainstream for tourism through announcements/promotional activities of the events. No doubt, there is a possibility of the emergence of negative externalities in the form of excess in the use of resources, cases of overtourism and overcrowding; and sometimes, the problem of unsuccessful planning and failure to implement or to attract the expected tourists to destinations. However, these issues can be tackled with the involvement of all the stakeholders taking note of the loopholes associated therein as well as the implementation of policy at national/local levels including sources of funding mechanisms to meet all the contingencies. In this regard, this chapter highlighted the role of finance in bringing the sustainability of sports tourism as an important sector for meeting the sustainable development goals of countries. Herewith, it is the role of the government and policy authorities to find out the factors/determinants for promotion and streamlining the sports tourism as an important segment of economic growth. It has been observed that sports financing in India needs a holistic approach specifically in terms of promoting sports events, which have cultural and heritage values. There will be a list of demand-led or demand-following factors on the one side; and on the other side, there will be a list of supply-led or supply-leading factors, which are relatively significant and need appraisal from the regional and institutional context. The underlying factors in developed and developing countries may not be the same, given the variation in their institutional, economic, and political framework. Focus should be given toward the developing and emerging economies wherein despite their voluminous scope for organizing sports tourism, they still lag behind from destination tourism as funding and financing in these areas generated a lacuna in possible destinations.

KEYWORDS

- **financing**
- **sports tourism marketing**
- **sustainable development**
- **technology**

REFERENCES

Bourdeau, P.; Corneloup, J.; Mao, P. Adventure Sports and Tourism in the French Mountains: Dynamics of Change and Challenges for Sustainable Development. *Curr. Iss. Tour.* **2002,** *5*, 22–32, https://doi.org/10.1080/13683500208667905.

Cho, H.; Joo, D.; Chi, C. G. Examining Nostalgia in Sport Tourism: The Case of US College Football Fans. *Tour. Manage. Perspect.* **2019,** *29*, 97–104. https://doi.org/10.1016/j.tmp.2018.11.002.

Funk, D. C.; James, J. The Psychological Continuum Model: A Conceptual Framework for Understanding an Individual's Psychological Connection to Sport. *Sport Manage. Rev.* **2001,** *4* (2), 119–150. https://doi.org/10.1016/S1441-3523 (01)70072-1.

Gratton, C.; Dobson, N.; Shibli, S. The Economic Importance of Major Sports Events: A Case Study of Six Events. *Manage. Leisure* **2000,** *5* (1), 17–28. https://doi.org/10.1080/136067100375713.

Gursoy, D.; Kendall, K. W. Hosting Mega Events: Modelling Locals' Support. *Annals of Tourism Research*, **2000,** 33 (3), 603–623. https://doi.org/10.1016/j.annals.2006.01.005

Higham, J. Sport Tourism as an Attraction for Managing Seasonality. *Sport Soc.* **2005,** *8*, 238–262. https://doi.org/10.1080/17430430500087419.

Hinch, T. D.; Higham, J. E. Sport Tourism: A Framework for Research. *Int. J. Tour. Res.* **2001,** *3*, 45–58.

Kaplanidou, K.; Karadakis, K. Understanding the Legacies of a Host Olympic City: The Case of the 2010 Vancouver Olympic Games. *Sport Market. Quart.* **2010,** *19* (2), 110-117.

Kim, C.; Kaplanidou, K. The Effect of Sport Involvement on Support for Mega Sport Events: Why Does It Matter? *Sustainability* **2019,** *11* (20), 5687. https://doi.org/10.3390/su11205687.

Lopex-Bonilla, L. M.; Reyes-Rodriguez, M. C.; Lopez-Bonill, J. M. Golf Tourism and Sustainability: Content Analysis and Directions for Future Research. *Sustainability*. **2020,** *12* (9), 3616. https://doi.org/10.3390/su12093616.

Pouder, R. W.; Clark, J. D.; Fenich, G. An Exploratory Study of How Destination Marketing Organizations Pursue the Sports Tourism Market. *J. Destin. Market. Manage.* **2018,** *9*, 184–193. https://doi.org/10.1016/j.jdmm.2018.01.005.

Shukla, P.; Shamurailatpam, S. D. Harnessing Information and Communication Technology Led Interventions for Mapping Overtourism: Prospects and Retrospect. In *Overtourism, Technology Solutions and Decimated Destinations*; Hassan, A., Sharma, A., Eds.; Springer: Singapore, 2021. https://doi.org/10.1007/978-981-16-2474-2_12

Taleghani, G. R.; Ghafary, A. Providing a Management Model for the Development of Sports Tourism. *Procedia- Soc. Behav. Sci.* **2014,** *120* (19), 289–298. https://doi.org/10.1016/j.sbspro.2014.02.106.

UNDP United Nations Development Programme: Annual Report, 2018. https://sustainabledevelopment.un.org/content/documents/20470VNR_final_approved_version.pdf.

UNTWO Sport Tourism and the Sustainable Development Goals (SDGs), 2019. https://www.e-unwto.org/doi/epdf/10.18111/9789284419661.

UNTWO Tourism Investment Report: Global FDI Greenfield Investment Trends in Tourism, 2020. https://webunwto.s3.eu-west-1.amazonaws.com/s3fs-public/2020-09/FDI-Tourism-Report-2020.pdf.

Waitt, G. Social Impacts of the Sydney Olympics. *Ann. Tour. Res.* **2003,** *30* (1), 194–215. https://doi.org/10.1016/S0160-7383 (02)00050-6.

Weed, M.; Bull, C. *Sports Tourism: Participants, Policy and Providers*; Elsevier Butterworth-Heinemann: Burlington, 2004.

Wilson, R. The Economic Impact of Local Sport Events: Significant, Limited or Otherwise? A Case Study of Four Swimming Events. *Manage. Leisure* **2006,** *11* (1), 57–70. https://doi.org/10.1080/13606710500445718.

Yang, J. J.; Lo, H. W.; Chao, C. S.; Shen, C. C.; Yang, C. C. Establishing a Sustainable Sports Tourism Evaluation Framework with a Hybrid Multi-Criteria Decision-Making Model to Explore Potential Sports Tourism Attractions in Taiwan. *Sustainability* **2020,** *12* (4), 1673. https://doi.org/10.3390/su12041673.

UNTWO. Tourism Statistics. www.unwto.org/tourism-statistics-data.

UNTWO. World Tourism Network Organization. https://www.unwto.org.

CHAPTER 14

The Place and Reflections of Financial, Economic, and Climate Change in Tourism Policies

HALIDE BILGIN[1], ENKELEDA LULAJ[2], and HAVANA KUM[3]

[1]*Institution of Social Sciences, Anatole University, Eskisehir Turkey*

[2]*Faculty of Management in Tourism Hospitality and Environment, Haxhi Zeka University, Peja, Kosovo*

[3]*Institution of Social Sciences, Anatole University, Eskisehir Turkey*

ABSTRACT

Climate change can be defined as a change in the climate caused by human activities which results in the loss of the natural environment and affects the economic and financial development of the country. Climate change poses threat to all living things, countries, and different sectors of businesses due to the consumption of fossil fuels and increase in greenhouse gas, which at the same time affects the financial development of the country. The purpose of this study is to determine the role of financial and climate change in the existing plans and policies. In line with this goal, the policies made for climate change in Turkey and the economic and financial development in different countries have been reviewed. Tourism is one of the most important sectors that contributes to the development of all countries, influencing the growth of financial and economic development. As a result of rising climate change, tourism sector has become most affected. For this purpose, tourism policies are reviewed, published publications are scanned and analyzed. As a result of present studies on climate and financial change, tourism industry

Dynamics of the Tourism Industry: Post-Pandemic and Post-Disaster Perspectives and Strategies.
Gül Erkol Bayram and Anukrati Sharma (Eds.)

will begin to have an impact on the future, as the negative and positive outcomes of climate change on the tourism industry, which is one of the victims of climate change, may also contribute to this process. Reduction and formation of greenhouse gas mitigation actions should be taken by the tourism sector and adaptation measures should be identified to reduce the effects of climate change. Also, at the end of the study, it was found that although policies have been developed a lot, the issue of climate change has not been given much space, and tourism strongly influences the growth of economic and financial development of countries.

14.1 INTRODUCTION

Tourism and the environment have a common relationship. Environment is a source of tourism for the surrounding regions. In order for this relationship to continue smoothly, the environment must exist. It contains all the factors that will affect living beings, inanimate beings, and the actions of living beings. The decisiveness of the relationship between tourism and the environment is also effective in the economic development of the neighboring regions, but it also leads to environmental degradation. The fact that these distortions occur is a dangerous element from the point of view of tourism. A tourism movement that is not used properly leads to the deterioration of the natural environment, a bad impact on the climate, and water, air, and soil pollution in a negative way. When these adversities are experienced, it shakes the ecological balance and brings climate degradation to the fore. Thus, climate change, which is one of the most prominent factors today, has become a major problem faced by all living things. Climate change causes deterioration of the health of living beings; the bad influence of human and animal generations causes the disorganization of the ecosystem. More human factors are at the root of climate change. For example, population growth, industrialization, pesticides greatly affect climate change. Greenhouse gases, melting of glaciers, rise in sea level, and garbage thrown into the environment are the main factors that trigger climate change. Increasing global warming also leads to climate change. In this context, climate change affects agriculture, energy, health, and tourism closely. It is known that the tourism sector creates an interaction in many areas, such as social, cultural, economic, political, health, and environment throughout the countries. However, many countries receive a significant revenue from the tourism sector. For this reason, the service sector plays a dominant role in economic development (Özekici and Silik, 2017:59). Climate and weather conditions also have an impact on tourism activities. For example, people can go to the sea, ski, or take part in a nature walk only when climatic conditions are appropriate and suitable for their activities. On the other hand, people living in countries that do not have suitable climatic conditions for vacation travel to other regions may choose to spend vacation in a place where they have favorable climatic conditions. (Polat et al., 2019: 241). Hence, it is clear that climate is one of the most important factors which increases the demand for tourist destinations. These climatic factors, such as heat, precipitation rate, temperature, humidity, and radiation play an important role in people's preferences. The resulting impacts have gone beyond being an environmental issue. The negative effects of climate change on the ecosystem, air, and water have disrupted the socioeconomic situation, affected human health, caused migrations, agricultural inefficiency, desertification. Climate change has caused many bad consequences. In short, every negative situation it creates then comes across as a secondary disaster.

Moreover, the problems that arise all over the world indicate that measures should be taken against this global disaster on a global scale (Akbulut, 2019: 1). In order for the first climate change event to have a more positive contribution to tourism, some plans, strategies, and policies are being developed. Such efforts make it easier to determine the impact of climate change on tourism and how to adapt to this effect. Especially in developing countries such as Turkey, the role of tourism in regional and national development is great. In the context of increasing competition, tourism, which is also a major source of income in the Turkish economy, may be adversely affected by these climatic conditions (Polat et al., 2019: 241). The impact of climate change on the tourism sectors of countries is most often observed in the context of ecotourism. Because many species of creatures are becoming extinct due to climate changes. According to the World Wide Fund for the Protection of Nature, nearly, 29 natural parks on Earth will be significantly damaged by climate changes. Therefore, the forms of travel to these areas may also be negatively affected by these conditions (Özekici and Silik, 2017: 67). The implementation of plans and policies at the national and international level will support the process of climate change in the long term. In other words, policies should be framed to adapt to climate change, to combat climate events, and to counter the positive or negative effects of climate. Here, the state should contribute to foreign policies as well as contribute to domestic policies (Kartal, 2018: 1). Many problems, such as changes in the amount of heat and precipitation, sudden hot and cold, floods, droughts, forest fires, population mobility, shake the climate change. For this reason, more emphasis should be placed on climate change in tourism policies. Tourism policy analyzes the research related to the tourism industry, tourist attractions, and the current state of the tourism sector (Erkol Bayram and Ak, 2018: 200). As for every state and every sector, planning is an indispensable tool for the tourism sector in terms of fulfilling its management functions (Erkol Bayram and Ak, 2018: 199). Hence, tourism policies are formulated every year and there are solutions to solve the existing problems. However, climate change has not been much thought about. Every plan and policy created for tourism also implements a successful tourism service. In this context, in the thesis study, it is aimed to examine the studies in the field of tourism policies of climate change to identify climate change, which is a current problem, and to determine its place in tourism policies. In the study, tourism policies were examined to find solutions to the problem of climate change. The reason why this topic is included in the thesis is that there has been little study on climate change in tourism policies. The fact that it is a rare subject when examined in the literature reveals the originality of the study. In this study, interview technique, which is one of the qualitative research methods, was used.

14.2 DEFINITION AND CONCEPT OF POLICY AND TOURISM POLICY

In order to achieve further development of tourism, it is necessary to plan appropriately within itself and with other sectors in order to give appropriate responses to people's demands on supply. The word policy is used in response to the meanings of politics and the ways, methods, and programs kept (Çevik, 2003: 136). Tribe (2010:7) refers to the general rules or guidelines that guide the behavior as a policy. According to Page and Connell (2014:226), the policy shows the goals and direction of an organization that it is desirable to follow for a specified period of time (Polat, 2016:26). In general, the concept of policy is defined as the work of influencing and directing state activities (Akay, 2009: 14) or as general guidelines for decision-making (Yağcı, 2007: 194; cited by Erkol Bayram and Ak, 2018: 201).

A policy is a method of selecting information obtained to determine current or future decisions and to guide or implement these decisions from among alternatives. Policies can be considered as a method of defense against threats that may occur in the development of the country. There are different definitions on the concept of politics. According to Şimşek

(2007:141), politics is "a measured and planned way to achieve a certain goal or to conduct business or government affairs." Can et al. (1999:148), on the other hand, define policies as general guidelines that guide decision-making (Polat, 2016: 26).

Demirkol and Cetin (2014: 24) define policy as objectives and the methods of any country or organization that must be followed to rsesolve an issue or achieve desired goals and determining the resources that will be used in regulating the organization of events and relationships both inside and outside the organization in the interests of the organization is the process where activities are maintained as long as it exists. Polat (2016: 26) defines it as "Politics is an endless action; given the changing environmental conditions, increasing knowledge, it is a process in which uninterrupted decisions can be reached and there must be social agreement." Tourism policy analyzes the research related to the tourism industry, infrastructure opportunities, tourist activities, tourist attractions, and the current state of the tourism sector" (Erkol Bayram and Ak, 2018: 200). Over time, the emerging concept of sustainable development for the protection of social, cultural, and natural resources has become the main responsibility of states. Along with this, policies for the social welfare of people have been started to be created. Tourism policies have been established with the measures taken by the states in order to ensure the inflow of foreign currency and the creation of jobs, revenues that have increased with the increasing growth of tourism. Governments, political circles, sectoral tourism organizations, entrepreneurs, and local governments are trying to implement the tourism policies they have developed within the scope of their areas of responsibility (Erkol Bayram and Ak, 2018: 200). In short, tourism is initially considered within the scope of national policies due to its economic contributions to national income, foreign exchange inputs, employment, and therefore national well-being. Later, tourism, its economic contribution, as well as the protection and promotion of cultural and natural values to create a bridge between countries and to contribute to world peace can take place in the context of national and international policies with like attributes (Coşkun, 2004: 1). The creation of a successful and strong tourism policy depends on the support of local people, technological developments, informed tourism authorities, employees, and implementations.

Tourism policy, as a guide to concrete activities and economic policy, aims to give you some direction in economic science as a branch of abstract format for viewing (Timur, 1998: 316) develops the tourism sector in a country, steer, and control by governments determined using a variety of tools in order to approach the target; Tourism policy is the totality of actions and interventions

(Usta, 2014: 193). In other words, if a series of regulations, rules, guidelines, and instructions are the primary targets in developing tourism policy, the daily activities of a tourism destination and tourism affect the long-term development of collective and individual strategies (Tuncel, 2019: 60). When the literature is examined, different definitions of tourism policy are observed. Definitions of tourism policy are as follows: Tourism policy is the way of managing and acting on plans prepared for the development or support of tourism in a country, region, city or destination and covering a certain period. The basic principle of any tourism policy is that a country must ensure that its economic and social contributions to tourism will benefit as much as possible. The main purpose of a tourism policy is to improve the progress of the nation and the lives of its citizens (Tuncel, 2019: 60).

Tourism policy is defined differently by the authorities. For example, the Ministry of Tourism and Promotion of Turkey, established in 1963, has defined its tourism policy as "a dynamic policy aimed at maximizing the economic, social, and cultural benefits provided in tourism and minimizing costs" (Cihan Ağbay, 2019: 42). According to another definition, tourism policy is "To all the people of a country to provide opportunities for physical and spiritual relaxation by participating in tourism while protecting the environment for their needs in the most appropriate sub- and superstructure of public administration in the field of tourism with the aim to establish directly or indirectly any intervention" (Cihan Agbay, 2019: 42). According to Biederman et al. (2008: 460), tourism policy refers to the direction of a particular country, region, environment or a separate destination to take action when developing tourism. The important point for tourism policy here is that the country or region can make maximum use of the economic and social contributions of tourism. The ultimate goal of the tourism policy is to improve the progress of the lives of the country or region and its citizens (Cihan Agbay, 2019: 42).

14.3 THE CONCEPT AND SCOPE OF TOURISM PLANNING AND POLICY

Planning is a kind of policy making and decision-making, but it consists more of common and interrelated decisions than individual decisions. Each planning has a specific purpose; it is to prepare action plans for the predicted future and to implement these plans. There are different definitions on the concept of planning in the literature. According to Page and Connell (2014: 225), planning gives the state the opportunity to determine which areas and sectors can be used for development purposes by exchanging views with other stakeholders. According to Cook et al. (1999: 235) planning expresses that it includes predicting the future, determining the goals for the results to be achieved in the future, as well as determining and implementing the actions necessary to achieve these goals (Polat, 2006: 32). Içöz et al. (2009: 78) define planning as follows: "Planning is itself a process of identifying goals and ways to achieve them that may change over time depending on changes in the environment, as well as developing a hierarchy of plans." According to Demirkol and Çetin (2014: 30), planning is the determination of the necessary works to be done in order to achieve predetermined goals and the ways to be followed in the process of fulfilling these works (Polat, 2006: 32).

The main concept of economic activities is planning, and planning should be in harmony with the tourism industry. Tourism is the most growing, income-generating sector in the world against the backdrop of economic stagnation. Hence, well-prepared and organized planning strategies are needed. The different factors included in tourism make it mandatory to implement planning-based policies.. These factors include the scope and complex structure of tourism, the positive and negative effects of tourism, accommodation, and environmental protection. (Altınay et al., 1995: 87).

If we define tourism as temporary trips and stays of people from where they live all the time to somewhere else, the fact of active planning arises in the face of this issue. Apart from the positive aspects of tourism, threats to the tourism industry have started to increase with the rise of negative aspects. In order to reduce these negative effects, a plan and practice should be managed. The planning role of the state in tourism is especially prominent in supply and demand planning. At this point, the state can play the role of an arbitrator between the parties whose interest dec in conflict. For example, the involvement of local governments in tourism planning and coordination, especially in order to maximize tourism development, is also a reason why tourism should not be left only to private sector entrepreneurs for profit (Ağbay, 2019: 81).

In tourism planning, for a certain period, the goals to be achieved in the tourism sector, tools to be used to achieve these goals, things to do, monetary and physical facilities, business by whom, when, how soon may not be realized (Erkol Bayram and Ak, 2018: 204). Tourism planning is the most important stage in the development and progress of tourism. The tourism plan shows how, when and by whom, with what resources the visions and strategies are created for the development of tourism, and the ability of countries to generate high income will be made. If tourism is developed in an unplanned manner, there will be a sharp drop in the number of tourists, leading to a decrease in the income. Tourism regions that develop in a planned way have more competitiveness. It is observed that the effects of tourism are increasing in the regions where tourism activities are developing.

When an unplanned development is encountered because of a lot of density in the territory, the carrying capacity of the destination will be badly affected environmentally and socially. If necessary measures and precautions are not taken, various consequences, such as irreparable nature destruction, distorted and unplanned urbanization, waste and destruction of resources, and sociocultural problems may arise (Atça Tonbil, 2019: 23). For this reason, plans should be implemented to avoid negative consequences, such as cultural, social, economic, environmental and always developable tourism. According to another definition of the tourism plan, the aim of this plan is to determine the optimal goals in the tourism sector, the resources to be used to achieve these goals, the measures to be taken, and to establish a realistic alignment between the goal-source-measure based on analysis according to the country's economic system, political regime (Alaca, 1997: 38). According to Içöz et al. (2009: 91), tourism planning "is a continuous cycle process based on scientific research consisting of an index of coordinated guidance activities between the decisions and actions of organizations that will guide the tourism system and ensure that the system follows the dec given direction. Polat (2016: 48) defines it as "Tourism development plan indicates general planning. According to Inskeep (1988), tourism planning consists of research, preparation, identification of targets, survey, analysis and synthesis, policy and planning, implementation and monitoring, including a seven-step process (Polat, 2016: 48). Burkart and Medlik (1988: 70) express the growth and development of tourism in a country or region of may cause the waste of resources because of land use, the demands of visitors and the conflicts of local community, seasonal nature of tourism, seasonal unemployment. The sources of degradation that will result in a healthy economy based on one or two industries would lead to some problems which emphasizes the importance of planning (Polat, 2016: 48).

A poorly managed tourism industry leads to economic disruptions, a rapid decrease in income, social and cultural pollution, and the destruction of natural resources. For this reason, in order to maximize tourism activity and minimize the damages that may occur, permanent policies should be made in the tourism sector and measures should be taken in a planned and programmed manner and they should be implemented. Tourism has many purposes. Tourism policy also determines the ways to achieve these goals and strives to maximize the economic, social, and cultural profits from tourism. In order to fulfill these goals, good planning is needed. A document published by the Ministry of Tourism and Promotion also defines tourism policy as follows: it is a dynamic policy aimed at maximizing economic, social, and cultural gains from tourism and minimizing costs (Yeğen et al., 2011: 241). The tourism policy has its own characteristics. These are given as follows:

1. Tourism policy is dynamic: The reason for the dynamics is the organic connection between it and the tourism economy. The reason is that when a change occurs in the tourism economy, it is correct to state that a change occurs in the content and objectives of the tourism policy. For example, an economy that accepts travel for health within the scope of tourism includes the health sector in its tourism policies and the creation of policies that encourage health travel.
2. The tourism policy is multifaceted: The tourism event has a complex structure as it includes many components. This is due to the fact that the social aspect of tourism is excessive and has a solid cooperation with various factors. According to an example given by Alaca (1997: 21), there is a close relationship between tourism and economic growth, social development, and urbanization. Although natural resources are the creative element of tourism, there may be some consequences, such as tourism being a tool that destroys natural resources. For these reasons, a good tourism policy provides for some regulations to take advantage of nature, while on the other hand, it must cover measures that prevent the destruction of natural resources.
3. Tourism policy is institutional: It is the decision-making bodies of tourism in the direction of supply and demand bodies that make up the tourism policy. The decision-making bodies in the direction of supply are designated as tourism enterprises, while the demand bodies are the consumer group that makes up the tourism demand. In addition, public institutions are also among the decision-making bodies that affect supply and demand. All these constitute the tourism policy.
4. Tourism policy is rational: In order to achieve the intended goal in tourism policy, the resources, methods, and techniques to be used

in tourism must be based on rational principles. If this is not the case, the resources have been wasted and the contribution that the national economy expects from tourism may be negative in the long run. The presence of these features in tourism policies contributes to the tourism of the country or region. The tourism sector requires high investments. Therefore, tourism policies are needed.

14.4 CLIMATE CHANGE CONTEXT AND SOME CHALLANGES

We have been hearing a lot of news about floods, forest fires, high temperature rises that have been on the agenda lately. The world climate is also affected by these differences. In short, the world experiences a lot of rainfall in 1 year and drought the other year. Among the reasons for this situation, climate change is the most important factor (Aydemir and Şenerol, 2014: 3). When the literature is examined, different definitions about climate change are included. According to Aydemir and Şenerol (2014: 5); Climate change is defined as "statistically significant changes in the average state of the climate or its variability that last for decades or more." Climate change can occur due to natural internal processes and external forcing factors, as well as constant anthropogenic (human-caused) changes in the composition of the atmosphere or land use. Currently, climate change can be defined taking into account human activities that increase the greenhouse gas accumulations. For example, the United Nations Framework Convention on climate change (UNFCCC) says "in addition to natural climate change observed over comparable time period, directly or indirectly, human activity that disrupts the composition of the global atmosphere as a result of a change of climate"(Aydemir and Senerol, 2014: 5).

There are three basic concepts related to climate change. These concepts are the greenhouse effect, global warming, and climate change. The uneven greenhouse effect causes global warming and climate change over a long period of time. Global warming and climate change is a process that occurs as a result of the fact that greenhouse gases produced by humans beings as a result of their various activities, which accumulate intensively in the atmospheric layers near the earth, and terrestrial radiation is blocked by the greenhouse effect (Hırlak, 2020: 4). There are many reasons that cause climate change. The first of these is the phenomenon of increasing global warming, which can be given as an example. The change in climate is closely affecting the agriculture, industry, energy, health, and tourism sectors. The tourism sector offers visitors the beauties created by the climate. Types of tourism are also increasing in places where there is a favorable climate. But in regions where the adverse conditions of the climate are reflected, this diversity cannot show an increase. For example, the ability of people to ski, take advantage of the sea-sand-sun trio, and go hiking depends on the appropriate climatic conditions. Otherwise, people living in areas that do not have a suitable climate prefer places where the climatic conditions are beautiful more. As a result, weather and climate factors are two important variable factors for the tourist who demands the region (Polat et al., 2019: 241). As important as the weather and climate factor are, the issue of climate change also makes it important. Because, as a result of climate change, we are experiencing events such as changes in sea level and vegetation, differentiation of ecological life, and precipitation regime. In this case, the tourism sector may be badly hit for a whole year. Climate change, to be precise, will affect human life. Especially the service sector may be under more pressure. The tourism industry, which is included in the service sector, is one of the affected ones. This is due to the fact that the opportunities offered by the climate should be positive in tourist preferences and movements (Aydemir and Şenerol, 2014: 386). Throughout history, there is evidence of climate change effects on earth's crust. The climate has changed many times, especially with the melting of glaciers and movements at sea level. Many data reveal that climate change has a great impact on temperature, precipitation, water levels, and life. Due to climate change affecting temperature, it is predicted that warming will increase over the next two decades. Temperature increase has its effects on water resources, food, health, natural disasters, and biodiversity. In particular, events such as a change in climate and an increase in temperature also affect the precipitation that falls in the regions. In the case of insufficient rainfall, a decrease in water resources may be observed. Şenerol (2010: 31-31) proved that water is decreasing with the following information; according to the climatic conditions of 1990, the amount of water per capita in Turkey per year is 3,070 cubic meters. But most of this water is not available in places where water is needed. Even if we accept that the climatic conditions will not change, the amount of water per capita per year in Turkey in 2050 will be 1240 cubic meters only due to population growth. Considering that we will have a climate that is 10 times drier as a result of global climate change along with our increasing population, the amount of water per capita in Turkey per decum will be between 700 and 1910 cubic meters in 2050. This is the same amount of water per capita on the island of Cyprus at the moment. In other words, with its changing climate and increasing population, it is estimated that Turkey may become a thoroughly water-poor country in 2050. Another issue caused by climate change is biodiversity. In particular, the habitats of polar bears, penguins, fish, and destruction of microorganisms by changes in soil structures negatively affect the plant and animal species. Another effect of climate change is food production. As a result of the bad influence of the climate, it becomes difficult to grow crops (Şenerol, 2010: 27-31-32-34).

Climate change is a global problem that affects the lives of all living things on Earth. Various efforts have been taken against this problem. In addition, we see a lot of national and international effects of climate change. Mostly, agriculture, forests, water resources, sea level affect the life of living beings. Fossil fuels, which are frequently used in the formation of climate change, are effective in the rapidly growing population. The Earth's climate has changed in different sizes over the years. The fact that volcanic eruptions affect landforms is the cause of climate change, and also located in the high mountains and the Poles. Due to global warming and melting glaciers, sea level rises. Ocean and marine species that live in the depletion of the favorable environment, the emission of more carbon dioxide gas, environment and water pollution, increasing the change of day and night temperatures reflect the consequences of climate change and acid rain.

14.5 THE RELATIONSHIP BETWEEN CLIMATE CHANGE AND TOURISM

Climate has a relationship with many sectors. The tourism sector is one one such sector. In the tourism sector, tourists are offered the beauties of the climate, its attractiveness, and in this case, activities in accordance with the weather conditions are included. For example, in order to realize winter tourism, the snow potential must be high. People expect favorable climatic conditions to exist for their holidays, they do not prefer unsuitable climates. In particular, it is observed that natural factors are more effective in the decision-making processes of tourists. Natural factors also seem to affect coastal and winter tourism. The increase in temperature leads to algal growth and jellyfish infestation on the beaches. Due to warming climate and in the absence of snow, ski resorts will not be able to function and suffer without the income. Climate is the dominant factor in tourism. It not only provides tourists supply but also affects the tourism demand (Şenerol, 2010: 91).

There are many models deciphering the relationship between climate and tourism. The most commonly encountered is the Tourism Climate Index. The Tourism Climate index (TCI) was first introduced by Mieczkowski in 1985. Şenerol (2010: 91) described the Tourism Climate Index as follows. This concept is used for the purpose of determining the tourism quality average for the tourist in a mixed way using systematic climatic elements. Initially, TCI was determined with the help of the relevant 12 separate climate variables. Climate by reducing the number of variables to seven meteorological data (maximum daily temperature, daily average temperature, daily minimum relative humidity, daily average relative humidity, total rainfall, total gunesleme hours, and average wind speed) is reduced. These seven climate variables constitute five sub-indices of TCI. With the work of the United Nations (UN) in 1990, the UNFCCC was opened for signature at the UN Conference on Environment and Development held in Brazil in 1992, and the process of combating it was launched at the global level. In order to reduce the impact of climate change, the UNFCCC has aimed to reduce the impact of greenhouse gases, which are emitted due to human activities (Bayazıt, 2018: 222). Events such as floods, earthquakes, landslides, forest fires that have been occurring for many years have also played a role in changing the climate. Changing climatic conditions have left positive or negative effects. In this case, climate change is associated with tourism sectors.

The tourism sector is a major source of income for countries. The service sector plays a dominant role for developed and large countries. One of the most important factors affecting tourism activities is the climate. In this regard, the fact that the Mediterranean climate will become extremely hot after climate change may indicate that these regions are not preferred by tourists and travel expenses will shift to other regions. In addition, due to the impact of climate change on destinations, visitors will be able to be more selective in their destination decision-making time. It can be said that this process will increase the importance of marketing effectiveness in tourism activities (Özekici and Silik, 2017: 63). One of the facilities dealing with winter tourism that will be negatively affected by climate change is the ski resorts. Due to excessive warming of the regions, the snow layer weakens and the income of the ski resorts may decrease. These reasons damage the economy of the regions where the tourism sector is located (Özekici and Silik, 2017:64). Climate change affects ecotourism more in the tourism sector of countries. This is because many living things are becoming extinct due to climate change. A large part of tourism activities takes place depending on climatic conditions. One of the effects of climate change on tourism is especially due to airline travel, which is the main source of emissions. This provides an interesting example. Because changes in the position and intensity of jet streams will increase turbulences that will reduce comfort, while extending distances and creating costs (Demiroglu and Ülgen, 2018:

7). Climate change is the most important factor affecting human life. The tourism industry is also a service-based sector. External and natural factors play an important role in tourism movements and in the decision-making.. It is argued that according to many, there may be worse changes in the future. For this reason, not only the tourism sector but also every sector should take precautions against threats that may occur. Tourism is a constantly growing industry and is one of the best ways to get rid of the stress that people experience all year round. Climatic features determine the tourist season. In addition, climate change also gives priority to whether tourism will be long or short in that season. People especially expect that the climate of the destinations they will go for holidays and recreation should be plenty of sun with little precipitation and low humidity.

Climate change has been a factor affecting people, animals, ecological systems, and many sectors throughout history, and it is difficult to find a direct solution to the issue of climate change. Judging by the events that have happened in the past and predicting the change in the climate, there is no distant future for tourism. There are many reasons in the literature on climate change. Greenhouse gases, human activities, natural disasters trigger climate change. The tourism sector is at a level that can be affected by climate change and therefore some precautions need to be taken. In order to prevent climate change a little, carbon emission policies should be established for each country, but doing so is based on cooperation on a global scale. Furthermore, in order to create a society that is aware of the threats of climate change and its future conditions, local elements of attraction should be introduced through the media or social media (Özekici, 2017: 65). It is necessary to increase public awareness, to share accurate weather information, to create disaster plans and insurance guarantees and to ensure safety against storms that may occur. For high rainfall and flood disasters; it is important to build safe accommodation facilities and protect wetlands. For drought, it is necessary to store rainwater and purify sea water. Using more renewable energy sources is one of the measures that can be taken regarding climate change. For the impact of ecosystem and biodiversity, national parks and protected areas need to be monitored to improve biodiversity, ensure species survival, and identify new areas to protect or relocate species living in their natural habitat. If possible, create wildlife corridors. In order to keep the sea temperatures stable, it is necessary to grow temperature-resistant corals and to monitor the temperatures regularly. When climate change can greatly affect snowfall, alternative approaches should be considered and the continuity of winter tourism can be ensured by using artificial snow (Şenerol, 2010: 124). Ways to prevent climate change begin in humans. Of course, measures taken alone may not have a big impact, but even the slightest change is a hope for the future.

14.6 REFLECTIONS OF FINANCIAL AND ECONOMIC CHANGE IN TOURISM POLICIES

Economic and financial development in the tourism industry has been hit hard by the COVID-19 pandemic in all countries of the world. Depending on the duration of the crisis, revised scenarios based on country statistics show that the potential economic and financial shock could range between a 60 and 80% decline in the international tourism economy by 2020. Governments of various countries beyond immediate measures to support economic and financial development through tourism are also developing recovery strategies including removing travel restrictions, restoring traveler confidence and rethinking the tourism sector for the future (OECD, 2020: 75). According to a research conducted in 51 OECD countries, the great importance of the tourism economy in financial development is emphasized, including domestic and foreign tourists, employment enterprises, and domestic consumption (OECD, 2020: 387). According to the authors (Chang, Chia-Lin, Hui-Kuang Hsu, and Michael McAleer, 2017: 8), the findings emphasize that the use of market returns in the tourism stock sub-index as the sole indicator of the tourism sector, compared with the overall activity of economic variables on tourism stocks, is shown to provide an exaggerated and highly volatile explanation of tourism's financial conditions. The tourism industry generates growth in economic and financial development (Kasimati, 2015: 55-62). According to Pratt (2015), revenues from tourism show that tourism generates a large amount of economic activity, but the revenues that remain in destinations are often very small. Therefore, the use of economies of scale can be a way to maximize the benefits of tourism by influencing the financial development of countries. According to one research, it is emphasized that tourism, economic growth, and financial development are co-integrated because inbound tourism fosters economic growth in India both in the long run and in the short run. But it is suggested to promote policies to attract more international tourists (Ohlan, 2017: 9-22).

14.7 CONCLUSIONS

Environmental problems, especially due to increased greenhouse gas emissions, global warming, and climate change have affected the planet and have become a topic discussed all over the world. There have been various agreements and regulations on climate change that have been taken many steps at the international level. It is inconceivable that the tourism sector, which is in a very close relationship with the climate, will not be affected by the effects of climate change that will be experienced. The tourism sector will be affected by climate change both positively and negatively. When the effects are examined, the negative effects outweigh the positive effects. Therefore, in order to be at least affected by the negativity that will occur, great attention should be paid to compliance activities. More policies should be formulated and solutions to climate change should be found. It is also emphasized that tourism has a strong impact on the economic and financial development of countries through increasing the number of tourists, domestic and foreign. Therefore, as a policy, countries should create strategies to improve tourist attractions by increasing the economy and financial development because as stated in previous research, tourism, economic growth, and financial development are integrated with each other, as well as during COVID-19 pandemic nondevelopment of tourism has caused great damage to economic and financial development.

KEYWORDS

- **tourism industry**
- **pandemic**
- **disasters**
- **climate change**
- **recovery**
- **financial impact**
- **economic growth**

REFERENCES

Akça, Y. Türkiye'nin Kalkınma Planlarında Turizm Politikası. *Bartın Üniversitesi Dergisi* **2016,** 721–726.

Atça Tonbil, S. *Türkiye'de Turizm Politikası ve Planlaması*; İstanbul Üniversitesi Sosyal Bilimler Enstitüsü Yüksek Lisans Tezi: İstanbul, 2019; pp 4–53.

Aydemir, B.; Şenerol, H. İklim Değişikliği ve Türkiye Turizmine Etkileri: Delfi Anket Yöntemiyle Yapılan Bir Uygulama Çalışması. *Balıkesir Üniversitesi Sosyal Bilimler Dergisi* **2014,** 381.

Bayezit, S. İklim Değişikliği ve Turizm İlişkisinin Türkiye İç Turizmi Açısından İncelenmesi. *Anatolia: Turizm Araştırmaları Dergisi* **2018,** *29* (2), 221–231.

Chang, C. L.; Hui-Kuang, H.; Michael, M. A Tourism Financial Conditions Index for Tourism Finance, *Challenges*. **2017,** *8* (2), 23. https://doi.org/10.3390/challe8020023.

Cihan Aybay, N. *Türkiye Turizm Politikasının Oluşturulmasında Yönetimin Uygulanabilirliği Üzerine Bir Araştırma*; İnönü Üniversitesi Sosyal Bilimler Enstitüsü Yüksek Lisans Tezi: Türkiye, 2019; pp 41–107.

Çiftçi, F. *Türkiye'nin İktisat ve Turizm Politikaları İlişkisi*; Anadolu Üniversitesi Sosyal Bilimler Enstitüsü Yüksek Lisans Tezi: Türkiye, 2019; pp 23–92.

Erkol Bayram, G.; Ak S. *Turizmde Planlama ve Politika. Turizm Ekonomisi*; Değişim Yayınları: İstanbul, 2018; pp 200–214.

Hırlak, F. Yerel Düzeyde İklim Değişikliği ile Mücadelede İklim Eylem Planlarının Rolü ve İstanbul İklim Değişikliği Eylem Planı. Çanakkale On Sekiz Mart Üniversitesi Lisansüstü Eğitim Enstitüsü Yüksek Lisans Tezi: Türkiye, 2020.

Karakuş Kaçma, F. Bir Sosyal Politika Sorunu Olarak Küresel İklim Değişikliği: Türkiye Akdeniz İkliminde Bir Alan Araştırması; Gazi Üniversitesi Sosyal Bilimler Enstitüsü Doktora Tezi: Ankara, 2020.

Kasimati, E. Does Tourism Contribute Significantly to the Greek Economy? A Multiplier Analysis. *Eur. J. Tour. Hosp. Recreat.* **2016,** 7 (1), 55–62. https://doi.org/10.1515/ejthr-2016–0006.

OECD. *OECD Tourism Trends and Policies*; OECD Publishing: Paris, 2020. https://doi.org/10.1787/6b47b985-en.

OECD. Tourism Policy Responses to the Coronavirus (COVID-19). OECD, 1–75, 2020. https://read.oecd-ilibrary.org/.

Özekici, Y. K.; Silik, C. E. Türkiye'deki İklim Değişikliği Plan ve Politikalarında Turizm Sektörünün Yeri. *Gazi Üniversitesi, Turizm Fakültesi Dergisi* **2017,** 58–79.

Polat, S. *Türkiye'nin Turizm Politikalarının Beş Yıllık Kalkınma Planları ve Türkiye Turizm Stratejisi 2023 Kapsamında İncelenmesi*; Sakarya Üniversitesi Sosyal Bilimler Enstitüsü Yüksek Lisans Tezi: Türkiye, 2016.

Polat, E.; Düzgün, E.; Yeşiltaş, M. İklim değişikliğinin turizme etkisini belirlemeye yönelik hazırlanan lisansüstü tezlerin bibliyometrik profili. *Gümüşhane Üniversitesi Sosyal Bilimler Enstitüsü Elektronik Dergisi* **2019,** *10*, 240–249.

Ramphul, O. The Relationship Between Tourism, Financial Development and Economic Growth in India. *Future Busi. J.* **2017,** *3* (1), 9–22. https://doi.org/10.1016/j.fbj.2017.01.003.

Stephen, P. The Economic Impact of Tourism in SIDS. *Ann. Tour. Res.* **2015,** *52*, 148–160. https://doi.org/10.1016/j.annals.2015.03.005.

Şenerol, H. *İklim Değişikliği ve Türk Turizmine Etkisi Balıkesir Üniversitesi Sosyal Bilimler Enstitüsü*; Türkiye, 2020.

Yeğen, İ.; Aytar, O.; Erdemir, O. *Avrupa Birliği Turizm Politikası Bağlamında Türk Turizm Politikaları ve Karşılaştırılması*; Turizm Sempozyumu: Balıkesir, 2011.

CHAPTER 15

Putting the Tourism Industry Back on Its Feet Again: Argentina as a Main Case Study

MAXIMILIANO E. KORSTANJE

Departments of Economic, University of Palermo, Mario Bravo, Buenos Aires, Argentina

ABSTRACT

The specialized literature in disaster studies emphasizes the problems of applied research to forecast exactly the geographical zone or city where the disaster will take a hit. The term disaster appeals precisely to the lack of coherence in responses to outstanding, if not disturbing, events that impact notably society. For this reason, disaster management adopted the figure of resilience and adaptation to deal with post-disaster (pandemic) scenarios. Over the recent years, some voices have alerted on the importance of resilience in the tourism industry. The global pandemic that originated in Wuhan (China) rapidly disseminated throughout the world, paralyzing not only the tourism industry but also the global trade. The COVID-19 obliges us to rethink tourism in a feudalized (atomized) world without tourists. The current travel behavior, as well as the geographical borders, are being reformulated according to a post-pandemic situation. Beyond the material losses and lives, the COVID-19 seems to be a foundational event that redrawn the global geographies. This book chapter deals with the opportunities and challenges of tourism in Argentina just after the COVID-19. We look to answer the question: is post-disaster tourism literature an efficient instrument to put the activity back on its feet again?

Dynamics of the Tourism Industry: Post-Pandemic and Post-Disaster Perspectives and Strategies.
Gül Erkol Bayram and Anukrati Sharma (Eds.)

15.1 INTRODUCTION

The recently originated global pandemic, which is known as COVID-19 (SARSCOv2), has devastating economic effects and durable social effects in society. The tourism industry, of course, was not an exception (Donthu and Gustaffsson, 2020; Haywood 2020; Korstanje and George 2021). As Baum and Hai (2020) put it, the tourism industry was the main carrier of the pandemic, but paradoxically its main victim. The technological advances not only globalized the world but also connected overcrowded cities in hours. Hence, governments disposed of restrictive measures, which included the closure of borders and airspace without mentioning social distancing and lockdown to cut public circulation. International flights were suddenly canceled, affecting the industry as well as global trade as never before (Bakar and Rosbi, 2020; Foo et al., 2020; Wassler and Fan, 2021). Whilst some voices alerted on the urgency of adopting a multidisciplinary method to study tourism, above all in a world without tourists (Wen et al., 2020; Korstanje and George 2021), others emphasized COVID-19 as a rebirth towards a more sustainable industry (Prayag 2020; Gossling et al., 2020; Higgins-Desbiolles, 2020). Whatever the case may be, scholars agree that the timeframe of recovery for the industry, as well as other sub-service sectors, progresses at a snail's pace (Prideaux et al., 2020; Zhang et al., 2021). To some extent, recent studies have widely proved that the classic forecasting methodologies are simply obsolete to prevent global pandemics of the calibre of COVID-19 (Zhang et al., 2021; Sigala 2020; Carr 2020). New situations require more adaptative methods instead of precautionary measures. In tourism management, literature has abandoned the risk perception theory, which is a theoretical platform borrowed from social psychology just after 2001, to embrace enthusiastically new dominant paradigms. The term risk perception, doubtless, sets the pace for "adaptation" or resilience (Sigala 2020; Korstanje and George 2021; Fotiadis et al., 2021). In these terms, adaptation means the possibility of coordinating efforts to deal successfully with post-disaster scenarios instead of preventing them. The idea of resilience was originally presented by Victor Frankl to denote his traumatic experiences living in concentration camps during Nazi Germany. At a closer look, the term speaks to the social abilities developed by individuals or communities to mitigate or recover from adverse situations. In disaster-related studies, resilience was successfully borrowed by psychology as a buzzword that shed light on multiple facets and dimensions. Resilience not only makes communities stronger but also reinvigorates post-disaster recovery (Amaratunga and Haigh, 2011; Malalgoda et al., 2014). The capacity for resilience corresponds with the successful responses of society to disasters or chaotic situations (Miles, 2015). In the tourism and hospitality fields, scholars, and policymakers believed over years that the industry was strictly sensitive to disasters. Disaster not only harmed the organic image of the destination but accelerated its inevitable decline (Cavlek, 2006). These studies certainly overlooked the capacity of resilience of tourist destinations (Yang and Nair, 2014). Over recent years, some seminal works have turned the attention on resilience as a key factor that revitalizes obliterated zones and cities (Biggs et al., 2012; Robinson and Jarvie 2008; Mair et al., 2016). In a nutshell, the body of academic theoretical corpus overly says that resilience greases the rails of foreign investment, allowing a rapid recovery as well as synergy among all stakeholders (Portoff and Neal, 1994; Tucker et al., 2017; Seraphin et al., 2017, 2020). Having said this, the present book chapter explores the practical implications of the lockdown in Argentina as well as the recovery efforts by the tourism industry to re-activate local consumption. The book chapter explores not only the impact of COVID-19 in the tourism industry, with a focus in Argentina, but also the future guidelines for the industry to get back on its feet again.

15.2 COVID-19 AND ITS AFTERMATH IN TOURISM

In the past, the tourism industry has faced similar situations with virus outbreaks (Rassy and Smith, 2013; Monterrubio, 2010). At a closer look, like invisible risks, viruses, diseases, or even poor food and water supply are pondered as more dangerous by tourists than other palpable risks. This happens simply because tourists are reluctant to those dangers they cannot see or anticipate (Lepp and Gibson, 2003). This reminds us that those risks which remain inexpugnable to the human´s eye cause panic (Korstanje 2010; McKercher and Prideaux 2014). Earlier, interesting research has reviewed the consequences of different viruses on the organic image of the tourist destination (Solarin 2015; Chen et al., 2021). The Film Contagion (2011) directed by Steven Soderbergh and starred by Matt Damon, Jude Law, Gwyneth Paltrow, and Laurence Fishburne, alludes to the desperately adopted measures of governments to control a global disease. These measures comprise the closure of borders and cancelation of flights as well as the imposition of quarantines. The plot situates the origin of the pandemic in Hong Kong, China just after Beth Emhoff´s travel (Paltrow), which disseminates a mortal virus after she meets with a former lover in Chicago. In the film, the protocols are a replica of the steps authorities adopted in former real pandemics. Besides, the social discontent and the turmoil and lootings in the streets occupy a central position in the plot. To some extent, pandemics and chaos are inevitably entwined. To the economic losses the global pandemic generates, one must add the social costs of unemployment, higher levels of anxiety and frustration that are sometimes crystallized in public riots and lootings (Novelli et al., 2018; Ritchie and Jiang, 2019; Cooper, 2019; McKercher and Chon, 2004). For the sake of clarity, pandemics not only affect host guests' relations but engender chauvinist expressions directed overly against foreign tourists. This seems to be the case that is well-documented by Mostafanezhad, Cheer, and Sin (2020) about COVID-19 and the hostility against Chinese tourists. Per their outcome, the pandemic wakes up global anxieties, which were rechanneled against Asian tourists. Authors coin the term "the China problem" to exhibit the degree of intolerance expressed by users in travel blogs revolving around Chinese visitors. At a first glimpse, COVID-19 changes not only the geopolitical relations (tensions) but also the travel behaviour. In other cases, material asymmetries between classes, ethnicities, or human groups are brought into the foreground (Duro et al., 2021). As Francesc Romagosa (2020) eloquently observes, the tourism industry enters an unparalleled state of crisis, where the lack of planning and uncertainness prevail. Since nobody knows what will happen with COVID-19 in an immediate future, no less true seems to be that the real diagnosis of the impact on the industry is based on mere speculations. What is more important, scholars acknowledge that COVID-19 represents a founding event that changed the industry as never before. In this line, similar policies orchestrated by health authorities in earlier virus outbreaks—like in Ebola, H1N1, or SARS—are not sufficient in the case of Coronavirus. One of the immediate effects, which are triggered by pandemics, associates with domestic (proximity) tourism. In a recently published book, entitles *The nature and future of tourism,* Korstanje and George (2021) argue that the pandemic ushered the global capitalist system into an inevitable fragmentation. Nation-states are facing internal turmoil while some provinces struggle to gain independence from the central authority. Because of this mayhem, the global world has been replaced by a feudalized world. Tourists who in the past were esteemed as ambassadors of democracy and economic prosperity—many of them coming from the First World—are not demonized and exiled as "undesired guest". Hospitality, which is the symbolic touchstone of Western civilization is in decline. Echoing Urry´s formula the tourist gaze, Korstanje and George formulate a novel neologism which is self-explanatory of the situation we live in: *the wicked-gaze*. This word evinces the hostility of locals against foreign tourists. In the new normal, the world is not only fractured, but the "other" is also undesirable. This raises a more than interesting question: is tourism recovery feasible after COVID-19?

15.3 TOURISM RECOVERY AFTER COVID-19: CHALLENGES AND OPPORTUNITIES

In numbers, the pandemic has left serious economic losses and victims worldwide. To date, the virus has infected more than 164,000,000 million people, while 3,413,168 has perished. The most affected countries are the US (600,951 victims), followed by Brazil (436,862), India (283,276), Mexico (220,489), the UK (127,691), and Italy (124,497) (Worldometers, COVID 19- cyphers on May 18, 2021). The economy is not pretty different. According to reports of the United Nations, the pandemic has led more than 34 million people into extreme poverty and pauperism. In the same way, the economy is forecast to lose almost $ 8.5 trillion in output over the next 2 years, whipping all gains earned in the previous 4 years. This marks the worse economic contraction since 1930. For 2020, the report estimates that GDP will fall 5% in developed economies. As stated in the introductory section, the different dispositions to restrict international travels, as well as the lockdown and social distancing, grind the tourism industry to a halt. The UNWTO estimates that international flights slumped from 60 to 80% during 2020 (in comparison to 2019) (UNWTO, 2020). In this token, Fotiadis et al. report that the drop-in travel arrivals will continue within the next 12 months, which represents a serious blow for the economy of developed countries. At a closer look, this pandemic—far from earlier others—seems to be one of the deadliest outbreaks in contemporary history. COVID-19 left serious impacts on travel and tourism globally. The problem should be tackled off from two combined perspectives. From a business perspective, policymakers and Academia should work hard in adopting post-recovery steps. The mass inoculation of hotel and tourism staff is part of these policies. From a policy-making perspective, experts should start with a financial bailout, grants, and fiscal aid to facilitate the rapid recovery of the sector. All the more astounding, those countries that depend on international tourism go through devastating effects because of COVID-19. To keep back to the same profit levels as the pre-pandemic contexts, a stable reopening not only seems not to be advisable but also unfeasible. The authors conclude that the economic losses (of almost 50% of profits) will remain until the next summer in 2022 (which means almost 15 months). Therefore, governmental support is vital for the industry to get back on its feet. In addition, as experts agree, a safe vaccine massively inoculated is the only element that placates the nefarious consequences of the pandemic. In consonance with this, Rastegar et al. (2021) lament that the restrictive travel bans associated with lockdowns are eroding not only social ties but also ushering the lower socio-economic classes into extreme poverty. COVID-19 exposed the material asymmetries of global capitalism, making some sectors more vulnerable than ever before. Quite aside from the fact that poverty has been triplicated in recent months, the pandemic allows making a just tourism future. Authors coin the term *just recovery* to signal the process (mainly based on distributive justice) that helps to correct the already-existing injustices of the system. In this token, an efficient recovery measure depends strictly on how local rights, interests and benefits are coherently articulated.

To cut the long story short, a new ethical dimension should be situated at the center of recovery strategies for rethinking tourism as well as its derived local imbalances. Lastly, A. Elnasr Sobaih et al. (2021) revised the role of small enterprises in the recovery facet post COVID-19. With a focus on Greater Cairo, Egypt, researchers found that restaurant owner-managers have shown more resilience to the pandemic than their hotel counterparts. In this way, it is safe to admit that effects on the tourism industry vary depending on enterprise type and subsector. Although the government started a program to give tax cuts for tourist organization, reducing the costs of energy while absorbing the full worker's salaries, small tourism enterprises are characterized by informal practices as well as unregulated financial resources, which make them particularly more vulnerable. Without external support, small enterprises will inevitably perish. These researchers distinguish two types of organizational resilience dimensions: planned and adaptive. While the former refers to the precautionary programs oriented to prevent the consequences of the disaster, the latter signals recovery strategies adopted in the post-disaster conjuncture. Since small companies do not often have recovery plans, they should be supported by governments. Surprisingly, their findings reveal that small enterprises are not only more adaptable (sustainable) but also more resilient to COVID-19 because of their lower cost operations. These outcomes are in sync with other interesting publications and applied research that is mainly associated with entrepreneurship, governance, and sustainability (Yeh, 2020; Pinos Navarrete and Shaw, 2021; Dias et al., 2021; Picarozzi et al., 2021).

15.4 THE RECOVERY OF TOURISM IN ARGENTINA

To date (on May 20, 2021), the number of infected persons in Argentina is 3,371,508 while there are almost 71,771 victims. Argentina reaches the position of eleventh in the ranking of the most affected countries. The country is geographically located in the Southern area of South America, bordering Chile (West), Uruguay (East), Brazil, Paraguay, and Bolivia (North). Doubtless, Argentina is the largest Spanish-speaking country in the world and the second-largest country in South America, second only to Brazil. The country is divided into 23 provinces and the capital, which keep strict autonomy and their constitutions from the central administration. The country claims sovereignty over the Malvinas-Falkland Islands since 1930, a conflict revived just after a war with the UK happened in 1982.

As the previous backdrop indicated, the quarantine began exactly on March 19, 2020 when President Alberto Fernandez disposed of several nation-wide restrictions, which included the closure of borders and airspace as well as the cancelation of international flights and airports, followed by strict lockdown or restrictions on public circulation. The schools and universities were closed down while experts worked hard to robust the health system. Oscillating in different levels, the lockdown lasted until October 25 when the country passed from ASPO (aislamiento social preventivo y obligatorio—mandatory and preventive social isolation to DISPO (Distanciamiento social preventivo y Obligatorio—mandatory and preventive social distancing). Officially, the first days of January 2021 open the doors to the new summer (holiday) seasons as well as the vaccine campaign. Unfortunately, each province maintained a certain degree of independence while respecting the decisions of the central administration. This suggests that some provinces opt for stricter lockdowns (like Formosa) even to the extent of closing domestic borders, while others (like Buenos Aires city) follow more lax measures. The coordination, among provinces, far from being standardized followed higher levels of chaos and uncertainty nationwide. Meanwhile, the Tourism Ministry implemented a set of measures that encompassed the gradual re-opening of national borders and international outbound and inbound flights. Authorities made the pertinent arrangements to receive domestic and foreign tourists. The activities at domestic airports and flights were regularized according to strict new protocols. Though each province asked for different travel requirements, which included a PCR test, health coverage or insurance, and round-trip tickets, no less true was that the Tourism Ministry launched awards campaigns to bail out the devastating hotel sector as well as tour agencies. The different campaigns were accompanied by discounts and scoring programs to duplicate the purchasing power of passengers. Certifications such as travel safe or safe destination were issued by the Tourism Ministry to different tour operators or companies that met the health protocols against COVID-19. At the same time, program REPRO 2 (Recuperacion Productiva 2- Financial recovery program) was targeted to pay 10% of the wages of hotel and tourism staff during 2021. This measure is aimed at supporting the production of the tourism industry as well as supporting employment in the sub-service sectors. Economic activity notably increased during the successive months of January, February, March, and April. To boost local consumption and tourist activity, it is vital to add that the main successful strategy of the Ministry was centered on the national *campaign (compre ahora, disfrute despues—pay now but consume later).* It is important not to lose sight of the fact that the pre-travel campaign marked not only the end of lockdown but also encouraged citizens to visit the country first. The disposition was oriented to stimulate inter-state travel (or domestic tourism) while monitoring international tourism carefully. The pre-travel consisted of programs of presale designed in 2020 to give economic benefits to holders and tourism companies for 2021. As Korstanje and George (2022) document said that:

> *Registering on a specific website whilst loading the purchasing bills, tourists receive 50% plus in credits to be redeemed at the destinations once arrived. Visitors are encouraged to register and detail their personal information until 30 of October whereas a credit card issued by the Central bank allows tourists to buy for 50% of the incurred expenditures. To put a clear example, a family group buys—through a travel agency—a tour package which includes air tickets for 4 persons, hotel and excursion totalizing ARS 150,000. This prepaid tour is bound to Calafate city to be used in February of 2021. Meanwhile, two air tickets to Cordoba city are added to the package for ARS 10,000 to be redeemed in August of 2021. The invoices are loaded to the webpage of the Tourism ministry for a total of ARS 160,000, in which case the family receives a credit of ARS 80,000. Soliciting visitors receive a (virtual) electronic wallet for ARS 75,000 to be used in Calafate and ARS 5,000 in Cordoba since August. The economic program fostered by the Tourism Ministry not only seems not to be new—since it was applied worldwide—but also centres on the logic of paying now and consumption later. (Korstanje and George, 2022: 175).*

As early discussed, the tourism industry plays an ambiguous role as a vehicle towards economic recovery, but at the same time, it disseminates the virus. Those countries which adopted planned strategies aimed at reinforcing domestic tourism faced a rapid recovery facet which led to economic revitalization. However, far from experiencing the benefits of mass tourism in the pre-pandemic days, in post-COVID-19 tourism operates into a fractured world (most associated with the feudalization process), where a whole portion of workers are forced to stay at home. To put things in other terms, the multiplication of travel requirements accompanied by the rise of flight rates made tourism a benefit only circumscribed to a privileged elite. The new normal not only changed the travel behavior, forcing many impoverished Argentineans to stay at home, as well as the morphology of what geographers know as globalization but also affected the health situation in the months to come. The country faced a second wave of infection stronger and more contagious than the first one. The reopening temporarily gave oxygen to the tourism industry, but paradoxically, it not only aggravated the fragile health situation but also triplicated the cases during the next months, forcing President Fernandez to promote new restrictions.

15.5 CONCLUSIONS

The pandemic caused devastating effects for Argentina's tourism industry as well as other associated sub-service sectors. Policymakers estimate that almost 80% of hotels are on the brink of bankruptcy. Similar percentages apply for other subsectors as tour operators, local guides, commercial airline flights, rent-a-cars, and national parks. The present chapter further interrogated the pros and cons of the Argentinean government's decision to implement restrictive travel bans organized to stop the mass contagion. At the same time, the classic morphologies of travel have been radically shifted towards new forms of consumption. The program entitled as *pay now consume later* helped the tourism industry fasten the recovery timeframe but paradoxically laid the foundations for a new second wave. Because of this, the chapter devoted efforts to giving a fresh study based on the Argentina case that describes the new morphologies of travelling in a new normality. It provides evidence in two directions. On one hand, the present study case contributes directly to the incipient literature in the fields of recovery management in post COVID-19. On another, it provides insight into the atomization process, which today affects the industry. Ultimately, further investigation is at least needed for the real implications of COVID-19 in the industry and global trade.

As a foundational event, the multiplication of studies, books, and conferences taking the virus as the main object of study obscures more than it clarifies. The growth of literature saturates a chaotic stage (fragmentary) of knowledge production (Wassler and Fan 2021). It ushers in an undisciplined state of confusion and misleading diagnoses (Korstanje and George, 2021). How may we avoid speculative insights in tourism research just after COVID-19? How may we study tourism in a world precisely without tourists? Are tourists the only valid source of information we obtain in applied research? What are the relegated voices and actors within the system who would gain protagonism in a post-COVID society? And finally, is the so-called re-birth of tourism a new path towards a more sustainable industry? The above-formulated questions pose big challenges for tourism research in the not-so distant future. Even if we laid the foundations for a new understanding of the multiple real effects of the pandemic on travel behaviour, a future discussion is urgently needed.

KEYWORDS

- **resilience**
- **Argentina**
- **COVID-19**
- **tourism**
- **post-disaster tourism**

REFERENCES

Amaratunga, D.; Haigh, R. *Post-Disaster Reconstruction of the Built Environment: Rebuilding for Resilience*; John Wiley & Sons: New York, 2011.

Bakar, N. A.; Rosbi, S. Effect of Coronavirus Disease (COVID-19) to Tourism Industry. *Int. J. Adv. Eng. Res. Sci.* **2020,** *7* (4), 189–193.

Baum, T.; Hai, N. T. T. Hospitality, Tourism, Human Rights and the Impact of COVID-19. *Int. J. Contemp. Hosp. Manage.* **2020,** *32* (7), 2397–2407

Biggs, D.; Hall, C. M.; Stoeckl, N. The Resilience of Formal and Informal Tourism Enterprises to Disasters: Reef Tourism in Phuket, Thailand. *J. Sustain. Tour.* **2012,** *20* (5), 645–665.

Cavlek, N. *Tour Operators and Destination Safety*; Routledge: Abingdon, 2006; pp 338–355.

Carr, A. COVID-19, Indigenous Peoples and Tourism: A View from New Zealand. *Tour. Geogr.* **2020,** *22* (3), 491–502.
Chen, S.; Law, R.; Zhang, M., Review of research on tourism-related diseases. *Asia Pac. J. Tour. Res.* **2021,** *26* (1), 44–58.
Contagion. *Dir Steven Soderbergh. English*; 106 Minutes: Warner Bros Pictures, 2011.
Cooper, M. Japanese Tourism and the SARS Epidemic of 2003. In *Tourism Crises*; Routledge: Abingdon, 2017; pp 129–144.
Dias, Á. L.; Silva, R.; Patuleia, M.; Estêvão, J.; González-Rodríguez, M. R. Selecting Lifestyle Entrepreneurship Recovery Strategies: A Response to the COVID-19 Pandemic. *Tour. Hosp. Res.* **2021,** ahead of print, 1467358421990724.
Donthu, N.; Gustafsson, A. Effects of COVID-19 on Business and Research. *J. Busi. Res.* **2020,** *117*, 284–290.
Duro, J. A.; Perez-Laborda, A.; Turrion-Prats, J.; Fernández, M. COVID-19 and Tourism Vulnerability. *Tour. Manage. Perspect.* **2021,** 100819.
Foo, L. P.; Chin, M. Y.; Tan, K. L.; Phuah, K. T. The Impact of COVID-19 on Tourism Industry in Malaysia. *Curr. Iss. Tour.* **2020,** 1–5.
Fotiadis, A.; Polyzos, S.; Huan, T. C. T. The Good, the Bad and the Ugly on COVID-19 Tourism Recovery. *Ann. Tour. Res.* **2021,** *87*, 103117.
Gössling, S.; Scott, D.; Hall, C. M. Pandemics, Tourism and Global Change: A Rapid Assessment of COVID-19. *J. Sustain. Tour.* **2020,** *29* (1), 1–20.
Haywood, K. M. A Post-COVID Future: Tourism Community Re-Imagined and Enabled. *Tour. Geogr.* **2020,** *22* (3), 599–609.
Higgins-Desbiolles, F. Socialising Tourism for Social and Ecological Justice After COVID-19. *Tour. Geogr.* **2020,** *22* (3), 610–623.
Korstanje, M. Role of Mass-Media in Swine Flu Outbreak in Buenos Aires. *Anatolia* **2010,** *21* (1), 169–173.
Korstanje, M.; George, B. *The Nature and Future of Tourism: A Post Covid 19 Context;* Apple Academic Press: Palm Beach, 2021.
Korstanje, M.; George, B. *Mobility and Globalisation in the Aftermath COVID-19: Emerging New Geographies in a Locked World*; Palgrave Macmillan: Basingstoke, 2022.
Malalgoda, C.; Amaratunga, D.; Haigh, R. Challenges in Creating a Disaster Resilient Built Environment. *Procedia Econ. Finance* **2014,** *18*, 736–744.
Mair, J.; Ritchie, B. W.; Walters, G. Towards a Research Agenda for Post-Disaster and Post-Crisis Recovery Strategies for Tourist Destinations: A Narrative Review. *Curr. Iss. Tour.* **2016,** *19* (1), 1–26.
McKercher, B.; Chon, K. The Over-Reaction to SARS and the Collapse of Asian Tourism. *Ann. Tour. Res.* **2004,** *31* (3), 716.
McKercher, B.; Prideaux, B.Academic Myths of Tourism. *Ann. Tour. Res.* **2014,** *46*, 16–28.
Miles, S. B. Foundations of Community Disaster Resilience: Well-Being, Identity, Services, and Capitals. *Environ. Hazards* **2015,** *14* (2), 103–121.
Monterrubio, J. C. Short-Term Economic Impacts of Influenza A (H1N1) and Government Reaction on the Mexican Tourism Industry: An Analysis of the Media. *Int. J. Tour. Policy* **2010,** *3* (1), 1–15.
Mostafanezhad, M.; Cheer, J. M.; Sin, H. L. Geopolitical Anxieties of Tourism: (Im) Mobilities of the COVID-19 Pandemic. *Dialogues Human Geogr.* **2020,** *10* (2), 182–186.
Novelli, M.; Burgess, L. G.; Jones, A.; Ritchie, B. W. 'No Ebola… Still Doomed'–The Ebola-Induced Tourism Crisis. *Ann. Tour. Res.*, **2018,** *70*, 76–87.

Lepp, A.; Gibson, H. Tourist Roles, Perceived Risk and International Tourism. *Ann. Tour. Res.* **2003,** *30* (3), 606–624.

Piccarozzi, M.; Silvestri, C.; Morganti, P. COVID-19 in Management Studies: A Systematic Literature Review. *Sustainability* **2021,** *13* (7), 3791–3804.

Pinos Navarrete, A.; Shaw, G. Spa Tourism Opportunities as Strategic Sector in Aiding Recovery from COVID-19: The Spanish Model. *Tour. Hosp. Res.* **2021,** *21* (2), 245–250.

Prayag, G.Time for reset? COVID-19 and Tourism Resilience. *Tour. Rev. Int.* **2020,** *24* (2–3), 179–184.

Pottorff, S. M.; Neal, D. D. M. Marketing Implications for Post-Disaster Tourism Destinations. *J. Travel Tour. Market.* **1994,** *3* (1), 115–122.

Prideaux, B.; Thompson, M.; Pabel, A. Lessons from COVID-19 Can Prepare Global Tourism for the Economic Transformation Needed to Combat Climate Change. *Tour. Geogr.* **2020,** *22* (3), 667–678.

Rastegar, R..; Higgins-Desbiolles, F.; Ruhanen, L. COVID-19 and a Justice Framework to Guide Tourism Recovery. *Ann. Tour. Res.* **2021,** 103161.

Rassy, D.; Smith, R. D. The Economic Impact of H1N1 on Mexico's Tourist and Pork Sectors. *Health Econ.* **2013,** *22* (7), 824–834.

Ritchie, B. W.; Jiang, Y. A Review of Research on Tourism Risk, Crisis and Disaster Management: Launching the Annals of Tourism Research Curated Collection on Tourism Risk, Crisis and Disaster Management. *Ann. Tour. Res.* **2019,** *79*, 102812.

Robinson, L.; Jarvie, J. K. Post-Disaster Community Tourism Recovery: The Tsunami and Arugam Bay, Sri Lanka. *Disasters* **2008,** *32* (4), 631–645.

Romagosa, F. The COVID-19 Crisis: Opportunities for Sustainable and Proximity Tourism. *Tour. Geogr.* **2020,** *22* (3), 690–694.

Seraphin, H.; Korstanje, M.; Gowreesunkar, V. Diaspora and Ambidextrous Management of Tourism in Post-Colonial, Post-Conflict and Post-Disaster Destinations. *J. Tour. Cult. Change* **2020,** *18* (2), 113–132.

Séraphin, H.; Butcher, J.; Korstanje, M. Challenging the Negative Images of Haiti at a Pre-Visit Stage Using Visual Online Learning Materials. *J. Policy Res. Tour. Leisure Events* **2017,** *9* (2), 169–181.

Sigala, M. Tourism and COVID-19: Impacts and Implications for Advancing and Resetting Industry and Research. *J. Busi. Res.* **2020,** *117*, 312–321.

Sobaih, A. E. E.; Elshaer, I.; Hasanein, A. M.; Abdelaziz, A. S. Responses to COVID-19: The Role of Performance in the Relationship Between Small Hospitality Enterprises' Resilience and Sustainable Tourism Development. *Int. J. Hosp. Manage.* **2021,** *94*, 102824.

Solarin, S. A. September 11 Attacks, H1N1 Influenza, Global Financial Crisis and Tourist Arrivals in Sarawak. *Anatolia* **2015,** *26* (2), 298–300.

Tucker, H.; Shelton, E. J.; Bae, H. Post-Disaster Tourism: Towards a Tourism of Transition. *Tour. Stud.* **2017,** *17* (3), 306–327.

UN REPORT. COVID 19 to Slash Global Economic Output by 8.5 Trillion Over the Next Two Years. https://www.un.org/development/desa/en/news/policy/wesp-mid-2020-report.html

UNWTO. International Tourist Number Could Fall 60–80% in 2020. 2020. https://www.unwto.org/news/COVID-19-international-tourist-numbers

Urry, J. *The Tourist Gaze*; Sage: London, 2002.

Wassler, P.; Fan, D. X. A Tale of Four Futures: Tourism Academia and COVID-19. *Tour. Manage. Perspect.* **2021,** *38*, 100818.

Wen, J.; Wang, W.; Kozak, M.; Liu, X.; Hou, H. Many Brains Are Better Than One: The Importance of Interdisciplinary Studies on COVID-19 in and Beyond Tourism. *Tour. Recreat. Res.* **2020,** Ahead to print, 1–4.

Worldometers, COVID-19 Coronavirous Pandemic. https://www.worldometers.info/coronavirus/?utm_campaign=homeAdUOA?Si%3Ca%20href=

Yang, C. L.; Nair, V. Risk Perception Study in Tourism: Are We Really Measuring Perceived Risk? *Procedia-Soc. Behav. Sci.* **2014,** *144*, 322–327.

Yeh, S. S. Tourism Recovery Strategy Against COVID-19 Pandemic. *Tour. Recreat. Res.* **2020,** ahead of print, 1–7.

Zhang, H.; Song, H.; Wen, L.; Liu, C. Forecasting Tourism Recovery Amid COVID-19. *Ann. Tour. Res.* **2021,** *87*, head of print, 103149.

CHAPTER 16

Future Strategies in Tourism Destinations After Disasters and Epidemics

ALI TURAN BAYRAM, MIRAY KÖSE, and GÖKÇENUR AVCI

University of Sinop, School of Tourism and Hospitality, Department of Guiding, Sinop, Türkiye

ABSTRACT

In this study, primarily the importance of destination management and how disasters and epidemics affect destinations are mentioned. It is possible to talk about many factors that affect the number of tourists visiting a destination. Disasters and epidemics, which are among these factors, cause sometimes short-term and sometimes long-term decreases in the number of tourists. Recently, it has been seen that these factors have a say in the attractiveness of a destination. As a result of the success shown in destination management after disasters and epidemics, the impact of material and moral damage to the region will be minimized, perhaps it will be able to gain a new attraction (ethically debatable) with tourism types such as disaster tourism or dark tourism. The management of these processes in the destination can determine the future of the destination. In this context, the study offers suggestions for the methods that can be followed in order to minimize the effects of material and moral damages caused by disasters and epidemics in destinations.

16.1 INTRODUCTION

As one of the most important components of tourism, the most difficult concept to manage and market in terms of tourism is the destination. The

Dynamics of the Tourism Industry: Post-Pandemic and Post-Disaster Perspectives and Strategies.
Gül Erkol Bayram and Anukrati Sharma (Eds.)

concept of destination in tourism is intertwined with sustainability in a way and expresses the management of tourism development. Because in order for tourism to develop in an area, a planned and programmed approach should be applied. Managing destinations is seen as one of the most important issues for the tourism industry. The tourism industry stands out as a risky industry facing many crises. These crises can take place in various ways, sometimes they occur as disasters or epidemics and affect destinations. In order to keep this effect at its lowest level, good destination planning is a must. In this context, it is important to strengthen the tourism industry by including studies on disasters and epidemics scenarios or searching for ways out after disasters and epidemics in destination plans. The tourism industry can try to regulate the position of the destination in tourism after disaster and epidemic in different ways. Measures can be taken to quickly erase the traces of a disaster or epidemic, or they will be able to take a new initiative to turn the disaster or epidemic into an opportunity. The concepts of dark tourism or disaster tourism in the literature can be seen as the greatest evidence of this.

16.2 DESTINATION AND DESTINATION MANAGEMENT

The concept of destination can be defined in its simplest form as a place with touristic attractions or a place to go. When the destination is considered as a complex structure consisting of tourism attractions, tourism businesses, and local people, it turns out that it is quite difficult to manage (Özdemir, 2007). The concept of destination in tourism is seen as a concept intertwined with sustainability. The most important reason for this is that the development takes place with a planned and programmed approach. Considering the three pillars of the destination structure, consisting of business, attractions, and local people, it will be very useful to carry out destination management from all branches. Although the presence of attractive factors (both physical elements such as natural resources and monuments and social factors such as spoken language and the friendliness of local people) is a necessary condition, it is not sufficient to transform a region into a tourism destination (Buckley, 1994; Ross and Ahola, 1991). The independent destination management system with appropriate competencies and experts carries out a coordination activity (Dávid and Tőzsér, 2009). At this point, the cooperation of all stakeholders is important for the realization of destination management. Collaboration is a key factor for sustainable growth across regions and industrial sectors. Tourism, one of the largest industries in the world, has been the subject of some of the most powerful innovations in recent years. (Ammirato et al., 2014). Crisis periods are also included in the strategies created by the stakeholders in destination management. In order to increase the quality of tourists' destination experiences, it is a necessity to raise awareness of the local people on all issues. In addition, educating local people about the responsibilities they will take on in times of crisis is such an important requirement. The traditional development model based on an outdated tourism supply chain model seems insufficient and unsustainable to support tourism destinations in strong and globalizing competition (Ammirato and Felicetti, 2014). In this context, plans for all scenarios that will take place within the tourism system should be realized with the participation of tourism stakeholders. Ignoring the stakeholders in this system may cause different problems, as well as make it difficult for the destination to get out of the crisis environment it is in. For example, Michalko et al. (2015) presented a different approach by mentioning the importance of disappointment management in order to prevent problems that may occur in the destination as a result of not meeting the expectations of the tourists in the destination. While this approach talks about the negative consequences of not meeting the expectations of tourists in destination management, it is difficult to imagine the negative consequences of not providing tourist safety or this environment of trust for tourists during disasters and epidemics.

16.3 IMPACT OF DISASTERS AND EPIDEMICS ON TOURISM

Since the beginning of human history, many disasters have occurred. Although most of these disasters are caused by natural causes, the number of direct and indirect man-made disasters is undeniable. Epidemics, on the other hand, occur after people's conscious or unconscious interventions in the ecological system. The borders that have disappeared with the effect of globalization, the increase in theinteraction between people, and the fact that countries are more dependent on each other through cooperation cause disasters and epidemics to affect more than one country. Disasters and epidemics undoubtedly affect tourism negatively. However, it is possible to talk about exceptional cases. Disasters cause a decrease in tourist interest in the destination; there are differences according to the type of event, but this interest never falls to zero. On the other hand, natural disasters and epidemics are thought to have a certain tourism value (Rucińska and Lechowicz, 2014).

One of the most important aspects affecting a person's participation in tourism activities is their perception of security. So much so that Maslow mentioned the importance of safety needs as the next step after physiological needs in the hierarchy of needs (Maslow, 1943). People cannot think too much about visiting a destination without ensuring their safety. In this regard, Payam and Selçuk (2017) state that despite all the technological advances for the benefit of humanity to promote travel and tourism, people have an increasing fear of security. Kôvári and Zimányi (2010) stated in their study that security, which is an indispensable condition for the travel and tourism sectors, gained great importance after the 1990s, but security decreased significantly with terrorist incidents, local wars, natural disasters, epidemics, and pandemics. In addition, it has been stated in the study that the travel and tourism industry has faced many negativities due to the reasons mentioned above and how fragile tourism and destinations are. For example, the earthquake that occurred in the Indian Ocean in 2004 and the tsunami that took place after it left nearly 5 million people homeless in 14 countries, including Indonesia, Thailand, India, Sri Lanka, Bangladesh, Myanmar, Malaysia, and the Maldives, and caused the deaths of around 230,000 people (www.trthaber.com). Many of the tourists who came to the region were among those who lost their lives. This negative impact has affected the destinations not only in terms of the tourism sector but also in areas such as transportation and infrastructure, and the destinations affected by this disaster have become an unsafe destination for tourists. Allocating trust for a destination is not a short-term and easy process.

For centuries, the world has been shaken by a new epidemic, disaster, and crisis in every new period. Thousands of people and dozens of destinations are deleted from the world. The losses caused by disasters and epidemics are always quite large, no matter how the new and developing worlds change. As a result of these disasters and epidemics, there is always a decrease in demand for destinations compared to previous years. Tourists, on the other hand, lose confidence in these regions, and the image of the destination is badly affected as a result of social problems. (Yılmaz, 2004). For example, the SARS epidemic, in which the first case was seen in 2002, caused a substantial decrease in the number of tourists coming to Taiwan. The epidemic caused the tourism sector to collapse not only in Taiwan but also in the Asian countries of China, Hong Kong, Singapore, and Vietnam (McAleer et al., 2010). In the early stages of the Corona Virus epidemic, which emerged in Wuhan, China, cancellations were made in reservation and travel planning. Since the number of cases has increased since March 2020, travel restrictions, visa cancellations, closures of borders and curfews have been announced (Meninno and Wolff, 2020). The COVID-19 pandemic, which is one of the biggest epidemics in human history, caused a 74% decrease in world tourism movements in 2020, while at the beginning of 2021, there were 88% decreases in many destinations compared to the pre-pandemic level (UNCTAD, 2021).

16.4 TOURISM STRATEGIES AFTER DISASTERS AND EPIDEMICS

Disasters and epidemics have created a crisis environment not only in tourism but also in every field. Events that cause such great psychological, physical, economic, and social damage must, of course, have a system that is prepared in advance, that is revealed at the time of the event, and that makes plans for the future. The correct crisis management system created by destinations is very important in minimizing the damage and impact during the disaster and epidemic processes (Bilgiçli and Yıldırgan, 2018). The better the crisis and disaster management plan of a destination, the faster the negative process will be overcome. In this context, it is necessary to talk about some methods and strategies that destinations can use to overcome these processes with the least damage.

The sharing of information by the authorities on national and international media platforms about destinations after disasters and epidemics is seen as one of the most important issues. The presentation of transparent information by the authorities will prevent disinformation, so the increase in negative perception will be prevented at some point. In this context, it is important that the disaster and post-epidemic communication plans are made separately, especially for tourists who purchase services from a business in the provision of tourist information. The roles of these businesses are clearly stated in the plans, and their participation in the process is ensured. However, since it will be a little more difficult to reach independent tourists in this system, necessary information should be provided at the border entrance gates, and the procedures to be followed in extraordinary situations should be specified.

The increase in disasters and epidemics has also caused some changes in tourist expectations and demands. In particular, the concept of security has come to the fore, and tourists have turned to safe areas or safe types of tourism. These effects were sometimes short-lived, with reductions in tourist travel to disaster- or epidemic-affected areas. In this context, especially with the COVID-19 pandemic, travels to tourism centers have decreased, and it has been seen that people tend to prefer more individualized and nature-based types of tourism. This shows that it can be seen as a necessity for a destination to diversify its products in order to be affected by disasters and epidemics at a lower rate. Electronic tourism and recreation applications, which emerged with the development of technology and were demonstrated by the pandemic process, will also have an indispensable place in the future strategies of the destination. The fact that digital technologies have a voice not only in business life but also in social life has brought about sectoral changes (Bayram and Kavlak, 2020). In the tourism sector, there are serious changes in this area and applications, such as virtual tourism can be perceived as safer, although they cannot provide a full tourism experience. Fear or restrictions caused by disasters and epidemic periods can be overcome in virtual environments. For this reason, although a region is mentioned when a destination is mentioned, it is thought that destinations should be moved to the virtual environment over time. Although it is seen as a distant future in terms of tourism activities, the virtual universe "metaverse", which is open to discussion and even property sales are increasing day by day, is one of the important issues that should be included in the future strategies of destinations.

Climate change and the increase in epidemics are undoubtedly seen as an indicator that the crises in the tourism sector will continue. Considering that disasters and epidemics will continue, these issues should be considered more seriously in destination management plans. It should be a necessity to transform the "safe tourism" practices that emerged with the Covid 19 pandemic into a routine for every destination and for every period, and to provide more detailed trainings such as disaster education, first aid, and hygiene education within the education plans and to all tourism stakeholders. It is important not only for education but also for the pre-determination of disaster or post-epidemic implementation plans, both for the local people and for transferring them to tourist groups, taking into account that they can happen at any time. These plans will also prevent the sector from working faster and especially qualified personnel from turning to different sectors.

As stated, it is seen that tourism movements do not completely end after disasters and epidemics; destinations tend to engage in activities that will revive tourism activities, and sometimes a disaster or epidemic is used as an opportunity. For example, after the SARS epidemic, Taiwan made attempts to gain the trust of tourists with the right crisis management strategy and image management, and after the epidemic that emerged in 2002, it managed to increase the number of tourists to over 3 million again in 2005 (Wang, 2009: 79). As another example, it is possible to talk about new types of tourism as an element that increases the motivation to travel to the disaster- or flood affected region. These types of tourism are called disaster tourism and dark tourism. Dark tourism is defined as tourists visiting places in a region where sad events such as war, disaster, mass death, or mass murder are experienced. Although disaster tourism is sometimes seen as synonymous with dark tourism or as a branch of dark tourism, the concept of disaster tourism mentioned in this study differs, especially in terms of timing. The concept of disaster tourism includes visits to a region before the effects of the disaster are felt. In other words, it includes the visit of tourists to the region while the war, earthquake, tsunami, or epidemics in the region and the destruction caused by them continue. For example, tourists traveling to Wuhan before the impact of the COVID-19 pandemic has passed or touristic trips are taking place while search and rescue efforts are still ongoing in the region after the tsunami in Indonesia. Since these two concepts are seen as types of tourism that will support the development of the region in terms of tourism after disasters and epidemics, there are issues that should be evaluated in future strategies. In particular, the concept of disaster tourism will have an important impact on the development of the region and healing the wounds, even if it is open to discussion from an ethical point of view. So much so that after the forest fires that took place in many regions of Turkey in 2021, tourism professionals in the Marmaris region called for tourists to come to the disaster area and asked for economic support (www.hurriyet.com.tr) and in fact, they indicated the importance of post-disaster tourism movements in this way.

16.5 CONCLUSIONS

Planning for crisis periods in destinations is a very important issue. The plans contain strategies that show how to get out of crisis situations. It is important to determine and implement these strategies correctly, to include all stakeholders in the process, and to deal with events from different perspectives. Tourism authorities should be given great duties and responsibilities in making these strategies, which can be applied in disaster and post-epidemic situations, functional. The tourism sector is undoubtedly one of the sectors most affected by disasters and epidemics. Strategies to be implemented become even more important as the effect is felt more, especially in destinations dependent on the tourism sector. After disasters and epidemics, strategies such as preventing disinformation, providing management unity, providing trainings, and ensuring stakeholder participation come to the fore. In addition to these strategies, issues such as dark tourism and disaster tourism, which are discussed ethically but are also seen as an application to turn the crisis into an opportunity, are among the strategies that cannot be ignored.

KEYWORDS

- **destination management**
- **emergency response plan**
- **dark tourism**
- **disaster tourism**

REFERENCES

Ammirato, S.; Felicetti, A. The Agritourism as a Means of Sustainable Development for Rural Communities: A Research from the Field. *Int. J. Interdisc. Environ. Stud.* **2014,** *8* (1), 17–29.

Ammirato, S.; Felicetti, A. M.; Della Gala, M. Tourism Destination Management: A Collaborative Approach. In: Collaborative Systems for Smart Networked Environments. PRO-VE 2014; Camarinha-Matos, L. M., Afsarmanesh, H., Eds.; *IFIP Advances in Information and Communication Technology,* **2014,** 434, 217–226.

Bayram, A. T.; Kavlak, H. T. *Rekreasyon ve Teknoloji.* In *Rekreasyon Disiplinlerarası Yaklaşım ve Örnek Olaylar*; Güneş, S. G., Varol, F., Eds.; Nobel Yayıncılık: Ankara, 2020.

Bilgiçli, İ.; Yıldırgan, R. Turizm işletmelerinde afet yönetim süreci. Alanya 4. Uluslararası Sosyal Beşeri ve İdari Bilimler Sempozyumu, 2018, 3–5 Mayıs Antalya.

Buckley, R. Framework for Ecotourism. *Ann. Tour. Res.* **1994,** *21* (3), 661–669.

Dávid, L.; Tőzsér, A. Destination Managament in Hungarian Tourism. *Appl. Stud. Agribusi. Com.* **2009,** *3* (5–6), 81–84.

https://www.hurriyet.com.tr/ekonomi/10-gundur-yanginlarla-mucadele-eden-bolgelerdeki-turizmciler-cagrida-bulundu-tatile-bekliyoruz-41868537) (Accessed 15 Oct 2022).

https://www.trthaber.com/haber/dunya/gezegeni-yerinden-oynatan-tsunaminin-uzerinden-17-yil-gecti-639062.html, (Accessed 15 Oct 2022).

Kôvári, I.; Zimányi, K. Safety and Security in the Age of Global Tourism (the Changing Role and Conception of Safety and Security in Tourism. *Appl. Stud. Agribusi. Com.* **2011,** *5* (3–4), 59–61.

Maslow, A. H. A theory of human motivation. *Psychological Review,* **1943,** 50 (4), 370–396. https://doi.org/10.1037/h0054346

McAleer, M.; Huang, B.; Kuo, H.; Chen, C.; Chang, C. An Econometric Analysis of SARS and Avian Flu on International Tourist Arrivals to Asia. *Environ. Modell. Softw.* **2010,** *25* (1), 100–106.

Meninno, R.; Wolff, G. As the Coronavirus Spreads, Can the EU Afford to Close Its Borders? 2020. https://voxeu.org/content/coronavirus-spreads-can-eu-afford-close-its-borders

Michalko, G.; Irimias, A. Dallen, J. T. Disappointment in Tourism: Perspectives on Tourism Destination Management. *Tour. Manage. Perspect.* **2015,** *16*, 85–91.

Özdemir, G. Destinasyon yönetimi ve pazarlama temelleri İzmir için bir destinasyon model önerisi. Yayımlanmamış Doktora Tezi. T. C. Dokuz Eylül Üniversitesi Sosyal Bilimler Enstitüsü, İzmir, 2007.

Payam, M. M.; Selçuk, N. Turizm güvenliğinin sağlanmasında özel güvenliğin rolü. 3. Uluslararası Türk Dünyası Turizm Sempozyumu, 2017, 20–22 Nisan 2017, Bişkek/Kırgızistan.

Ross, D. E.; Iso-Ahola, S. Sightseeing Tourists' Motivation and Satisfaction. *Ann. Tour. Res.* **1991,** *18* (2), 226–237.

Rucińska, D.; Lechowicz, M. Natural Hazard and Disaster Tourism. *Miscellanea Geographica—Regional Stud. Dev.* **2014,** *18* (1), 17–25.

UNCTAD, COVID-19 and Tourism an Update Embargo (Assessing the Economic Consequences). United Nations Conference on Trade and Development, 2021. https://unctad.org/system/files/official-document/ditcinf2021d3_en_0.pdf,A.D (Accessed 15 Oct 2022).

Wang, Y. S. The Impact of Crisis Events and Macroeconomic Activity on Taiwan's International Inbound Tourism Demand. *Tour. Manage.* **2021,** *30* (1), 75–82.

Yılmaz, Ö. D. Turizm işletmelerinde kriz yönetimi ve konaklama işletmeleri yöneticilerinin krizlere ilişkin yaklaşımlarına yönelik bir araştırma. Yayınlanmamış Yüksek Lisans Tezi. 2004, T. C. Dokuz Eylül Üniversitesi Sosyal Bilimler Enstitüsü, İzmir.

CHAPTER 17

Domestic Tourism Post-COVID-19

MARIA BOALER

Institute School of Management Studies, REVA University, Bangalore, Karnataka, India

ABSTRACT

The COVID-19 pandemic had impacted the global travel and tourism industry. In India, the total loss incurred by the travel and tourism is estimated to be Rs 5 lakh crore or $65.57 billion and the organized tourism sector has reported a loss of $25 billion (IBEF, 2021; *Times of India*, 2020) In this chapter, domestic tourism, its trends, and practices post COVID-19 are studied in context of India. Domestic tourism is defined as "tourism involving residents of one country traveling only within that country." Due to the pandemic, many developments have happened within the tourism sector, leading to many changes in tourist consumer behavior and business processes, which have led to new tourism trends. The aim of the study is to uncover patterns and methods in Indian domestic tourism that may be replicated in other nations.

17.1 INTRODUCTION

The COVID-19 or the Corona pandemic crisis began in Wuhan, China, on December 2019. The World Health Organization declared the first outbreak of the virus on January 20, 2020. The COVID-19 pandemic challenged the world in 2020 with no vaccine, limited medical facilities, nonpharmaceutical interventions like global travel restrictions

Dynamics of the Tourism Industry: Post-Pandemic and Post-Disaster Perspectives and Strategies.
Gül Erkol Bayram and Anukrati Sharma (Eds.)

and stay-at-home orders became a priority (Gössling et al., 2020) and the world had never experienced such a large and devastating economic and social impact (Naumov et al., 2020). The pandemic lockdown and restrictions at the border impacted all the sectors of the economy (Goodell, 2020), but tourism and hospitality were the worst hit (Gössling et al., 2020). Closed borders halted planned flights, while government-enforced quarantine restrictions severely curtailed travel possibilities. Many businesses had suffered significant losses in business due to low travel demand. However, the impact of the COVID-19 health disaster on the tourist sector is unprecedented in terms of the number of people affected and the number of countries involved (Yang et al., 2020).Governments began to consider measures to reestablish travel and re-establish economic prosperity as the spread of diseases halted (Shm et al., 2020). In order to revive the tourism industry, major steps, which includes extensive restructuring and adoption of new standards and norms, will have to be taken up (Galvani et al., 2020). The disease also altered people's perspectives, interests, and attitudes. Meanwhile, as a result of the pandemic, consumers' views, tastes, and attitudes toward travel have altered (Peters, 2020). As a result, organizations' business plans in the aftermath of the pandemic must necessarily account for changes in visitor behavior and needs (Brouder et al., 2020). As a result of the epidemic, consumers' views, tastes, and attitudes toward travel have changed drastically (Peters, 2020). As the pandemic impacted consumer behavior in general and tourism behavior in specific, organizations need to identify and study these changes in consumer behavior (Brouder et al., 2020). According to a global examination of government's policy initiatives of restrictions across different countries, the major restrictions had a direct influence on diverse tourism activities (UNCTAD, 2020).Closing of borders, prohibiting public gatherings, and closing various tourist sites and restrictions on transit between countries had a major impact on both domestic and also international tourism. To alleviate the crisis, governments have established tourism-specific programs, as well as economic policies. Several of these programs are geared on increasing domestic tourism, which is a recurring subject in travel and tourism in large countries.

COVID-19 had an impact on Indian tourism industry, which accounts for 6.8% of GDP and employs 8% of the population, according to the World Travel and Tourism Council (WTTC, 2020). The coronavirus pandemic has caused havoc on India's tourism and travel industries. According to a study in the *Economic Times*, the Coronavirus would put 3.8 crore people out of work in the tourist and hospitality industry (*Economic Times*, 2020). The whole value chain associated to travel and tourism lost nearly Rs 5 lakh crore, or $65.57 billion, with loss of $ 25 billion from organised tourism industry alone. As per FAITH, a policy federation of various organisations, COVID-19 is projected to cost the tourism and hospitality industry Rs 10 lakh billion. Tourism, which employs 8.75 crore people (12.75% of the total employed population in 2018–2019) and contributes 5.06% to India's GDP, has been halted by the strike (2016–2017).The *Economic Times* (*Economic Times*, 2021). Despite the fact that tourism activity of domestic tourism had increased from 1.05 billion to 1.85 billion from the year 2012 to 2018, the pandemic slowed its growth pace. The COVID-19 pandemic, which has impacted travel like no other event in history and cost the global tourism industry US $ 1.2 trillion in revenue due to 850 to 1.1 billion fewer tourist arrivals in 2020, could result in the loss or threat of 120 million direct tourism jobs, according to the World Tourism Organization (UNWTO, 2020). Even though the domestic tourist activity climbed by 10% between 2012 and 2018, from 1.05 billion to 1.85 billion, but with the continued outbreak, travel restrictions, and a nationwide shutdown, all this had a negative influence on growth. In the aftermath of the outbreak, there has been a spike in local tourism owing to overseas travel restrictions. According to KPMG, domestic tourism in India is predicted to touch 2.8 billion dollars by the year 2022 (KPMG, 2020) According to the World Tourism Organization (UNWTO, 2020) estimation by 2022, 120 million job losses might occur.

The COVID-19 epidemic, which had an unprecedented impact on travel, cost the global tourism industry $ 1.2 trillion in revenue by 2020 due to 850 to 1.1 billion fewer tourist arrivals (UNWTO, 2020). The tourist industry directly adds to GDP by earning foreign exchange, providing employment, and supporting many sorts of enterprises, all of which contribute to the overall growth of an area. At the same time, due to foreign travel limitations, there has been a surge in domestic tourism in the aftermath of the epidemic. While those long-awaited foreign vacations appear to be a long way off, the rich world of domestic travel appears to have taken on a new shape, with travelocal appearing to be the new trend. Hence, as a result of this, it is time for even India to develop a plan for restoring the ailing tourist industry, which has significant ties to the MSME. The number of foreign visitor arrivals (FTAs) in the first half of this year was a record high, according to the Ministry of Tourism's recent publication, "India Tourism Statistics."

17.2 DOMESTIC TOURISM

In Brazil, domestic tourism accounts for 94%, while it is 80% in Germany, Japan, Mexico, India, United Kingdom, China, and the United States; it is above 70% in Italy and France (UNWTO, 2020). Domestic tourists in Spain accounted for more than 70% of arrivals and 45% of the tourism-linked GDP (Arbul et al., 2021) As a result, it's vital to understand domestic travel, especially the factors that influence domestic tourist arrivals and spending, as well as domestic tourist behavior. (Ferrer-Rosell and Coenders, 2018).

Domestic tourism must be encouraged in order to work on tourism restoration. If individuals wanted to visit domestically in their own nation, domestic tourism would substantially boost the economy. In addition, the new government project "Dekho Apna Desh" invites travelers to visit 15 different places by the end of 2022 in order to win a prize. Once international travel begins, the chance is expected to be open to foreign tourists as well. Domestic visitors, for example, are often price-conscious and spend less. Itineraries created specifically for anticipated target travelers would provide considerable profits.

The tourist industry has lost a significant number of jobs since the lockdown measures were introduced in early 2020, according to a research report released by the National Council of Applied Economic Research in September 2021 (NCAER, 2021). The study gave an estimate that 14.5 million jobs were lost in the first quarter of 2020–2021, 5.2 million in quarter two, and 1.8 million in quarter three of 2020–2021, compared to a predicted 34.8 million (direct) jobs in 2019–2020. This is attributable to a drop in both domestic and foreign tourism. India's domestic tourist sector was able to rebound following the pandemic, thanks to a host of new trends and practises. Domestic tourism in India has recovered following the epidemic, thanks to a slew of new trends and practises. The Association of Domestic Tourist Operators of India (ADTOI) has requested that the Tourism Ministry, which belongs to India Government, has declared the year of 2021 as "domestic tourism year in India" so as to give direction and purpose to the entire travel industry, which is suffering because of the pandemic. The tourism and hospitality industry has expected a loss of Rs 10 lakh crore as a result of COVID-19. According to FAITH, the tourism sector, where 8.75 crore people are employed (12.75% of the total employed population in 2018–2019) and contributes 5.06% to India's GDP, has been halted by the lockdown and travel restrictions, which were imposed due to the pandemic. After the pandemic, India's domestic tourism industry was able to recover, thanks to a slew of new trends and practises (*Economic Times*, 2022).

17.3 NEW TRENDS POST-COVID-19

Domestic tourism is witnessing new trends like "Revenge Travel," "workcations," "travel bubble," and "travelocal." These trends predominantly surfaced during the pandemic (*Traveltrendstoday*, *TimesofIndia*, *Economictimes*).

Revenge Travel: Revenge Travel is a new term for a type of travel where people start traveling as they are fed up of staying at home during lockdown. The term "Revenge Travel" is a riff on the 1980s Chinese concept of "Revenge Spending" when the country saw increase in consumer spending after restrictions were removed. Post-pandemic a new trend has surfaced among Indian travelers called as "Revenge Travel" or "Revenge Tourism," this has presented a huge opportunity for the Indian hospitality industry. In December 2021, Goa surpassed pre-COVID-19 air passenger footfalls, as did religious towns like Tirupati, Shirdi, and others. The Indian Council of Medical Research (ICMR) has mentioned that revenge tourism is one of the main reasons for new COVID waves to occur, and that strict COVID protocols must be followed.

Workation: Workation or staycation or bleisure are also new terms used in domestic tourism trends. Long hours of working, communicating to all virtually, etc., made people rethink that instead of staying at home, step out and go to a place that is less polluted and surrounded by hills or sea and continue the work. This has led to the new holiday trend called as staycation, bleisure, or workation, where the term itself means work and vacation or business with leisure. More and more the idea of workation is catching up more with young people. People are merging work and play by reserving hotels in the same city or going out to farms to get away from it all. According to MakeMyTrip's Chief Operating Officer, there has been a significant increase in staycation searches to places within 300 kilometers of metro cities like Himachal Pradesh, Uttarakhand, Rajasthan, and Goa. The Telangana state of India also recently launched a contest called "win a free Staycation" on Twitter to improve their tourism.

Travel Bubbles: The FICCI Tourism Committee of India proposed that two or three Indian states collaborate regionally to create "travel bubbles" to encourage domestic tourism. To enhance domestic travel, the FICCI e-Tourism conference advocated interstate and regional bubbles. "Cooperation, coordination, and establishing synergies is necessary to overcome the current crisis," according to the summit.

Travelocal: Due to the pandemic, international trips were not possible, domestic travel took a new form and travelocal became the new trend. People started visiting places in and around their vicinity, also the places that can be reached by using their own vehicles, that is, bike or car.

Road Trips: As people feel safer using their own mode of transport, like their own car or bike, more and more people started traveling to their destination via road. The main reason for people to opt for road travel is because of the fear of taking public transport or fear of traveling with strangers or fear of falling sick.

India New Venue for Destination Wedding: Prepandemic, most of the Indians, whether they are celebrities or not, loved hosting grand weddings abroad. These weddings are termed as "destination weddings." Post pandemic, destination weddings are still happening, but more in India's domestic destinations like the heritages places, palaces, exotic islands, hill stations, etc. (Subhadra, 2020).

17.4 ANALYSIS OF CONSUMER EXPECTATIONS

The following are the findings of a study done by Booking.com, a digital travel assistant firm, on people's travel plans after COVID-19 (Prashasti-Awasthi, 2020.). There are some initiatives taken by the central and state governments. It is possible to express that these activities can be useful for private sector. The following show some central government initiatives (traveltrendstoday, 2022; Onlinetyari, 2021, Travelworld, 2020).

Bharat Parv: Every year, during the Republic Day celebrations, a big event called Bharat Parv is held with the goal of building patriotic fervor and showcasing our country's rich and diverse cultural variety. Due to the COVID epidemic, Bharat Parv 2021 will be held on a virtual platform from January 26 to 31, 2021, and will be coordinated by the Ministry of Tourism. States, UTs, and several Ministries will have their own theme pavilion on the virtual platform. Bharat Parv's main theme will also include Aatmanirbhar Bharat and Ek Bharat Shrestha Bharat (T3 NETWORK, 2021).

Dekho Apna Desh: The ministry of India Tourism has declared for 2020 as Dekno Apna Desh year. The main aim of this movement was to increase public awareness of the country's rich culture and heritage while also encouraging people to travel around the country. The Dekho Apna Desh webinar series, which began as a result of the COVID-19 pandemic which started in the aftermath of the COVID-19 pandemic and will end in March 2022, has not only highlighted the well-known, be it destinations, heritage, experiences, arts, and crafts, and dance forms, but it has also brought off-the-beaten-path and lesser-known tourism destinations, products, and experiences, such as "Chau dance," to greater prominence and awareness. What was most likely supposed to be a short-term and test campaign targeted at engaging and educating the tourist industry as well as travelers, morphed into a long-term and test campaign that has clocked over 113 webinars since its inception in April 2020 and is still going strong. The webinar series, on the other hand, is being watched in more than 60 countries.

The project named SAATHI: The Tourism Ministry has also launched the SAATHI project to realize some important issues for COVID-19 and after this process. This project was handled with the Institution of India Quality. This Council has some rights to make real-for-fine tourism activities for all tourism food and hotel managements. This council has many tourism management 6180 units approximately. All of managements have quality and hygiene certificates.

Paryatan Parv: It is initiative of the Union Government of India that any person who travels atleast 15 places of India by 2022 within a year would be rewarded by the Government of India.

The portal named "Stranded in India": Ministry of Tourism and Culture in India has launched an unit named with called "Stranded in India." This portal was started for helping and supporting tourists not to forget any information about their cancellation of flights. Many travelers used the portal to obtain information from state and union territory tourism agencies, as well as regional offices of the Ministry of Tourism.

The day for Yoga internationally: Another important ceremony being celebrated internationally is Yoga. To commemorate "International Yoga

Day," the Ministry of Tourism organized a series of activities centered on the topic "Yoga at Home & Yoga with Family" (IDY).

During the lockdown, the Ministry of Tourism has started photographing and videotaping significant tourism and cultural sites using Drone cameras to capture the best footage at a proper time when the pollution levels are at their lowest. Most of the states in India have taken several proactive steps to get their domestic tourism back on track. The following are some state-wise initiatives:

- The Karnataka Government had launched Tourism Policy 2020–2025 as part of the World Tourism Day on September 27, 2020 with the main objective to revive domestic tourism. The policy goal is, by 2025, to create over 10 lakh direct and indirect employment in the industry, as well as Rs 5000 crore in investments.
- Rajasthan: During the pandemic's low point, this royal state maintained contact with tourists across the world by offering virtual tours of some of its most popular attractions, including Umaid Bhawan Palace, Mehrangarh Fort, and Jaswant Thada. With travel returning to routine, Rajasthan is now developing new tourism opportunities. Bicycle excursions are one of them; the state tourist board encourages visitors to view the city while pedalling. Eléonore and Ophélie, two French friends, created the Cyclin' Jaipur initiative.
- Goa was the first state to open its borders for tourism post-pandemic as the state is heavily dependent on tourism for its economy.
- Gujarat state also was promoting domestic tourism with safety guidelines, creating short travel itineraries to announcing the new Heritage Tourism Policy 2020–2025 all during the pandemic.
- The Madhya Pradesh government has eliminated all travel restrictions inside the state, allowing people to visit local attractions that are worth seeing. Their "IntezaarAapka" campaign emphasizes the desire of travelers to get away of their homes with their families, for a short duration of time. Heres when weekend trips come in handy.

17.5 CONCLUSIONS

The use of touristic goods and services by the citizens of a country within their own country is called domestic tourism. Domestic tourism does not have a foreign exchange-generating effect; this effect occurs with foreign tourism. However, domestic tourism has a great impact on tourism activities within the country. Domestic tourism does not have a foreign exchange-generating effect, this effect occurs with foreign tourism. However, domestic tourism has a great impact on tourism activities within the country.

Domestic tourism will, therefore, gain in replacement for international travel as a result of these activities, but effective and standardized safety operating procedures will give it a boost. The present revival of COVID in India, as well as the fact that the wave is not gaining traction, should bring some relief. The positive news of immunizations on the way will instil confidence. People are gradually leaving their homes. Unless individuals gain confidence in traveling and leave their homes, which is happening slowly currently, the situation would not improve. Let us hope that the year 2024 will see a rebirth of the "Domestic Tourism Year," bringing good fortune to all industry stakeholders and the restoration of tourism.

KEYWORDS

- **domestic tourism**
- **Indian tourism**
- **new trends**
- **tourism policy and planning**

REFERENCES

Arbulú, I.; Razumova, M.; Rey-maquieira, J.; Sastre, F. Can Domestic Tourism Relieve the COVID-19 Tourist Industry Crisis ? The Case of Spain. *J. Destin. Market. Manage. 2021, 20* (March), 100568. https://doi.org/10.1016/j.jdmm.2021.100568

Brouder, P.; Teoh, S.; Salazar, N. B.; Mostafanezhad, M.; Pung, J. M.; Lapointe, D.; Higgins Desbiolles, F.; Haywood, M.; Hall, C. M.; Clausen, H. B. *Reflections and Discussions: Tourism Matters in the New Normal Post COVID-19*; Taylor & Francis, 2020. https://doi.org/10.1080/14616688.2020.1770325

Economic Times. FAITH Releases India Tourism Vision Document Envisaging Goals and Benchmarks for Tourism Till 2035, 2020.

Economic Times. Coronavirus Impact May Render 3.8 Crore People Jobless in Tourism, Hospitality Sector, 2020.

Galvani, A.; Lew, A. A.; Perez, M. S.; Galvani, A.; Lew, A. A.; Sotelo, M.; Covid-, P. COVID-19 Is Expanding Global Consciousness and the Sustainability of Travel and Tourism. *Tour. Geogr.* **2020,** *0* (0), 1–10. https://doi.org/10.1080/14616688.2020.1760924

Goodell, J. W. COVID-19 and Finance : Agendas for Future Research. *Finance Res. Lett. 2020*, *35* (March). https://doi.org/10.1016/j.frl.2020.101512

Gössling, S.; Scott, D.; Hall, C. M. Pandemics, tourism and global change:a rapid assessment of COVID-19. *J. Sustain. Tour.* **2020.**

https://economictimes.indiatimes.com/industry/services/travel/faith-releases-india-tourism-vision-document-envisaging-goals-and-benchmarks-for-tourism-till-2035/articleshow/89615911.cms

https://onlinetyari.com/latest-news-articles/bharat-parv-2021-organized-i106801.html https://www.entrepreneur.com/article/358725

https://timesofindia.indiatimes.com/business/india-business/travel-tourism-sector-likely-to-lose-rs-5-lakh-crore-due-to-COVID-19-crisis-report/articleshow/78015989.cms

https://travel.economictimes.indiatimes.com/news/associations/ficci-e-tourism-summit-proposes-inter-state-and-regional-bubbles-to-promote-domestic-travel/77240549

https://www.traveltrendstoday.in/news/india-tourism/item/10604-dekho-apna-desh-incredible-india-s-show-window

IBEF. *Tourism & Hospitality Industry in India*, 2021.

Jiang, Y. Effects of COVID-19 on Hotel Marketing and Management : A Perspective Article. *Int. J. Contemp. Hosp. Manage. 2020*, *32* (8), 2563–2573. https://doi.org/10.1108/IJCHM-03-2020-0237

KPMG. *Leisure and Travel*, 2020.

Naumov, N.; Varadzhakova, D.; Naydenov, A. *January 2020 Which Further Escalated to a Worldwide Global Pandemic Declared on 11*; Taylor & Francis. https://doi.org/10.1080/13032917.2020.1771742

NCAER. *India and the Coronavirus Pandemic: Economic Losses for Households Engaged in Tourism and Policies for Recovery*, 2020.

Peters, J. Visit People Tourism Recovery After Disaster. *KPPM Strategy*, April, 2020.

Prashasti-Awasthi. 70% of Indian Travellers Will Be More Price-Conscious While Planning Trips Post COVID-19: Survey. *Hindu Business Line*, 2020.

Shm, B.; Blanchard, K.; Ragupathy, D. Progress in Disaster Science Are We There Yet ? The Transition from Response to Recovery for the COVID-19 Pandemic. *Progress Disaster Sci.*, *7*, 100102. https://doi.org/10.1016/j.pdisas.2020.100102

Subhadra. 2020. https://www.treebo.com/blog/travel-trends-after-covid19.

T3 NETWORK. *Ministry of Tourism's Year End Review 2020*, 2021.

UNCTAD. *Impact of the Pandemic on Trade and Development*, 2020.

UNWTO. 2020. https://www.unwto.org/news/COVID-19-unwto-calls-on-tourism-to-be-part-of-recovery-plans.

WTTC *Latest Research from WTTC Shows a 50% Increase in Jobs at Risk in Travel & Tourism*, 2020.

Yang, Y.; Zhang, H.; Chen, X. Annals of Tourism Research Coronavirus Pandemic and Tourism : Dynamic Stochastic General Equilibrium Modeling of Infectious Disease Outbreak. *Ann. Tour. Res.* **2020,** *83* (March), 102913. https://doi.org/10.1016/j.annals.2020.102913

Index

E

F

H

I

J

L

M

U

V

W

Y